TO COOK A CONTINENT

Destructive
Extraction
and the
Climate Crisis
in Africa

Through the voices of the peoples of Africa and the global South, Pambazuka Press and Pambazuka News disseminate analysis and debate on the struggle for freedom and justice.

Pambazuka Press – www.pambazukapress.org

A Pan-African publisher of progressive books and DVDs on Africa and the global South that aim to stimulate discussion, analysis and engagement. Our publications address issues of human rights, social justice, advocacy, the politics of aid, development and international finance, women's rights, emerging powers and activism. They are primarily written by well-known African academics and activists. Most books are also available as ebooks.

Pambazuka News – www.pambazuka.org

The award-winning and influential electronic weekly newsletter providing a platform for progressive Pan-African perspectives on politics, development and global affairs. With more than 2,500 contributors across the continent and a readership of more than 660,000, Pambazuka News has become the indispensable source of authentic voices of Africa's social analysts and activists.

Pambazuka Press and Pambazuka News are published by Fahamu (www.fahamu.org)

TO COOK A CONTINENT

Destructive Extraction and the Climate Crisis in Africa

Nnimmo Bassey

Pambazuka Press
An imprint of Fahamu

Published 2012 by Pambazuka Press, an imprint of Fahamu
Cape Town, Dakar, Nairobi and Oxford
www.pambazukapress.org   www.fahamu.org   www.pambazuka.org

Fahamu, 2nd floor, 51 Cornmarket Street, Oxford OX1 3HA, UK
Fahamu Kenya, PO Box 47158, 00100 GPO, Nairobi, Kenya
Fahamu Senegal, 9 Cité Sonatel 2, BP 13083 Dakar Grand-Yoff,
Dakar, Senegal
Fahamu South Africa, c/o 19 Nerina Crescent, Fish Hoek,
7975 Cape Town, South Africa

British Library Cataloguing in Publication Data
A catalogue record for this book is available from the British Library

ISBN: 978-1-906387-53-2 paperback
ISBN: 978-1-906387-54-9 ebook – pdf
ISBN: 978-0-85749-087-2 ebook – epub
ISBN: 978-0-85749-088-9 ebook – Kindle

Manufactured on demand by Lightning Source

Contents

The colony's economy is not integrated into that of the nation as a whole. It is still organised in order to complete the economy of the different mother countries. Colonialism hardly ever exploits the whole of a country. It contents itself with bringing to light the natural resources, which it extracts, and exports to meet the needs of the mother country's industries, thereby allowing certain sectors of the colony to become relatively rich. But the rest of the colony follows its path of under-development and poverty, or at all events sinks into it more deeply. Immediately after independence, the nationals who live in the more prosperous regions realise their good luck, and show a primary and profound reaction in refusing to feed the other nationals. ... African unity, that vague formula, yet one to which the men and women of Africa were passionately attached, and whose operative value served to bring immense pressure to bear on colonialism, African unity takes off the mask, and crumbles into regionalism inside the hollow shell of nationality itself. The national bourgeoisie, since it is strung up to defend its immediate interests, and sees no farther than the end of its nose, reveals itself incapable of simply bringing national unity into being, or of building up the nation on a stable and productive basis. The national front which has forced colonialism to withdraw cracks up, and wastes the victory it has gained.

Frantz Fanon, *The Wretched of the Earth*

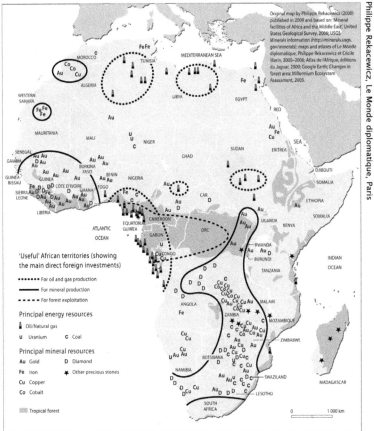

Original map by Philippe Rekacewicz (2000) published in 2009 and based on: 'Mineral facilities of Africa and the Middle East', United States Geological Survey, 2006; USGS Minerals information (http://minerals.usgs. gov/minerals); maps and atlases of Le Monde diplomatique, Philippe Rekacewicz et Cécile Marin, 2000–2006; Atlas de l'Afrique, éditions du Jaguar, 2000; Google Earth; Changes in forest area: Millennium Ecosystem Assessment, 2005.

Philippe Rekacewicz, Le Monde diplomatique, Paris

'Useful' African territories (showing the main direct foreign investments)

•••••••• For oil and gas production

———— For mineral production

— — — For forest exploitation

Principal energy resources

▮ Oil/Natural gas

U Uranium   C Coal

Principal mineral resources

Au Gold   D Diamond

Fe Iron   ✶ Other precious stones

Cu Copper

Co Cobalt

▒ Tropical forest

0   1 000 km

What the great powers covet

Preface

THERE ARE SOME in Africa who argue that having a valuable resource is not necessarily a curse. They say that nature's wealth is a blessing and that the curse happens only in relation to how resources are grabbed, owned, extracted and utilised. In other words, the curse is located firmly in the social structure of the world.

Let us start with a caveat about the word 'resource', which implies that nature's wealth is a bounty, ready for corporate robbery. But we as humans frame this dilemma of extraction incorrectly if we don't point out the intrinsic right of nature to survive on its own terms. Most importantly, we are part of Mother Earth, not apart from her. Her rights to exist and reproduce the conditions for all species' existence are not to be violated.

That said, everyone acknowledges that Africa is resource rich. That the continent has been a net supplier of energy and raw materials to the North is not in doubt. That the climate crisis confronting the world today is mainly rooted in the wealthy economies' abuse of fossil fuels, indigenous forests and global commercial agriculture is not in doubt. What has been obfuscated is how to respond to this reality. Indeed, the question peddled in policy circles is often what can be done *about* Africa. And, in moments of generosity, the question moves to what can be done *for* Africa.

This book looks at what has been done *to* Africa and how Africans and peoples of the world should respond for the collective good of all. The resource conflicts in Africa have been orchestrated by a history of greed and rapacious consumption. We ask the question: must these conflicts remain intractable? We will connect the drive for mindless extraction to the tightening noose of odious debt repayment and we will demand a fresh look at the accounting books, asking when environmental costs and other externalities are included: who really owes what to whom? Isn't Africa the creditor of the world, if we take seriously the North's 'ecological debt' to the South?

What makes possible the lack of regulation in Africa's extractive sectors, the open robbery and the incredibly destructive extractive activities? Leading the multiplicity of factors are unjust power relations that follow from and amplify the baggage of slavery, colonialism and neo-colonialism. From a Nigerian stand-point, but within the tradition of Pan-Africanist political economy and global political ecology, this book unpacks these issues and sets up bins for these needless and toxic loads.

Because of my own experiences, the pages that follow pay close attention to the oil industry in Africa, to the history of environmental justice struggles in the Niger Delta, to the discovery of oilfields in Uganda's Rift Valley, and to the big pull of the offshore finds in the Gulf of Guinea. As we examine the impacts of fossil fuel extraction on the continent, we also look at massive land grabs for the production of agrofuels and foods for export.

What can Africa do? And once our peoples decide, can the rest of the world act in solidarity? If not, will we continue on the path laid out by elites, a path that brings us ever closer to the brink? Must we live in denial even at a time of a rising tide of social and ecological disasters?

I cannot fail to express my thanks to Evelyn, my wife and comrade, who has remained a pillar of support in all my endeavours. I owe a huge debt of gratitude to Firoze Manji of Pambazuka Press and Pambazuka News, who challenged and spurred me to write this book and who patiently waited during its long gestation period.

I also owe very deep gratitude to Khadija Sharife and Patrick Bond, two of the most persistent opponents of exploitation and opacity working on our continent. Their collaboration on this project was most valuable, given their deep knowledge of the issues covered and readiness to share information as well as critique the work as it progressed. I cannot forget Bond going through the manuscript with two laptops bridged by flash drives as one battery ran down and he had to change to the other on our long bus and boat rides as we travelled to Yasuni from Quito. All I can say here is thank you for much inspiration and help. Needless to say, neither of them bears responsibility for what I say in the book.

Part 1
Unpacking Africa

Introduction: the pull of Africa

Placed on the slab
Slaughtered by the day
We are the living
Long sacrificed[1]

ONE OF THE worst gas flares in the Niger Delta is at a former Shell facility at Oben, on the border of Delta and Edo states. They have been roaring and crackling non-stop for over 30 years, since Shell first lit them. The flared gas comes from the crude oil extracted from the oil wells in the Oben field. As at more than 200 other flow stations across the Niger Delta, these gas flares belch toxic elements into the atmosphere, poisoning the environment and the people. Globally, gas flares pump about 400 million tonnes of carbon dioxide into the atmosphere annually. Here in Nigeria, the climate is brazenly assaulted both in the short term by gas flaring and over the long term because of the CO_2 emissions from this filthy practice. In the hierarchy of gas flares infamy, Nigeria is second only to Russia.

Gas flares and oil spills have attracted the attention of the world as the two most visible assaults on the Niger Delta. It was no surprise that when the Dutch parliament decided to hold a hearing on the activities of Shell in Nigeria, journalists and parliamentarians from the Netherlands decided to visit the region to see things for themselves.

I was at Oben on 18 December 2010 just after the United Nations' climate negotiations disaster at Cancun, accompanied by Sharon Gesthuisen, a Socialist Party member of the Dutch parliament, along with a Dutch diplomat and Sunny Ofehe of the Hope for the Niger Delta Campaign. Our journey started in Benin City early in the evening after the parliamentarian had flown in from Lagos. Escorted by a team of Oben community people, we set out on the hour-long ride along the highway from Benin City to Warri, a road noted for the high number of military checkpoints. They would make anyone think that Nigeria was at war. We meandered through the hazardous roadblocks made with trash hurled from nearby bushes and veered off the highway at Jesse, just before Sapele, from where we took a narrow winding road to Oben. Jesse is important in the tragic history of the Niger Delta: it was the community where a petrol pipeline fire killed about 1,000 poor villagers in 1998.

We got to Oben at about 7pm and were waved through a military checkpoint set up to guard the oil flow station and the belching dragons. Gaining entrance to the heavily guarded facility was easy; leaving was not. As soon as we arrived, a worker whom we happened across gave us a little talk about what went on there. People from the community complained about how they had had to put up with the flares for more than three decades while their dreams of jobs and development projects faded away.

The Dutch MP was amazed by what she saw. She was happy she had made this trip, otherwise she would have had to depend solely on the chaperoned visits arranged by the oil giant Shell in a bid to show how environmentally friendly they are. The flames leapt and roared relentlessly. We inched as close as we could before having to turn away because of the unbearable heat. As we turned to leave, the brightness of the village sky contrasted with the darkness of the homes that lacked electricity. But we could not leave.

Our cars were surrounded by soldiers of the Joint Task Force (JTF), a military force that became infamous when an armed unit was created specifically to punish the Ogoni people in the 1990s. The soldiers demanded to know by what authority we visited the gas flare site. They would not let us leave without producing an authorisation letter from the JTF headquarters. All our explanations that we were there at the invitation of the community fell on deaf ears. The presence of a Dutch parliamentarian as well as a diplomat meant nothing to these guys, who apparently knew their script. Hours went by. The darkness of the night struggled with the glow of the gas flares. The soldiers stuck to their guns.

The JTF men demanded our car keys and threatened to deflate the tyres. We would not leave the location that night, they insisted. Threats followed. Rifles were raised and then lowered. They would not call their superiors. They were the lords working at the behest of capital.

Eventually a Nigerian journalist who was on our team placed a call to the media relations officer of the JTF. After much foot dragging the soldiers wrote down the numbers of our cars and took our names, addresses and statements before letting us go at midnight. We rode back to Benin City in silence, each mulling over the hazards faced by communities living in the oilfields and the human rights abuses inflicted regularly on those who monitor or question the evils that go on in the land. To the Dutch parliamentarian, the events of the evening were a good introduction to the Niger Delta and the operations of the oil companies: exploit, degrade, abuse and punish the environment and the people. The scenario replays across the continent.

The African resource pull

The African continent has exerted a strong pull on the world for a long time. At first outsiders saw the continent as nothing more than a coastline. Beyond the coastline lay a dark, unknown land. While the land remained unknown to outsiders, kingdoms flourished on the continent and people lived in harmony with their environment in a cultural and spiritual relationship that held up a picture of sustainability that appears alien and exotic in today's world.

To Europeans, early knowledge of Africa was largely restricted to the Mediterranean fringes and later on to the southern tip, where from 1652 their foothold, established as a way-station for the India trade, left indelible marks. Even before then, slavery and looting were common. Arab merchants and their collaborators on the east coast were some of the first to see the black body as a resource to be extracted. But the English, Spanish, French and Portuguese colonialists took 20 million from the west and southern coasts.

When adventurers went deeper inland, the land yielded rich resources that astonished these so-called discoverers. Through the 18th century, Africa became the storehouse, with inexhaustible minerals, plant life and animals, as well as people.

Thomas Pakenham's *The Scramble for Africa* termed Africa a 'lottery': a winning ticket brought glittering prizes.[2] David Livingstone, one of the early adventurers, claimed that Africa could be saved through the tripod of commerce, Christianity and civilisation. Pakenham rightly responded that the invasion was four-legged and the fourth leg was conquest. He should have added that settler colonialism was a bizarre deviation from Christianity and civilisation, whereas commerce and conquest have persisted in a diversity of ways till today.

Often we read about how the invaders conned the continent with a bible in one hand and a musket in the other. As Archbishop Desmond Tutu has remarked, when the missionaries came to Africa, they had the bible and we had the land, but they asked us to close our eyes and pray, and when we opened them, we realised they now had the land while we had plenty of bibles. While we will not belabour that analysis, it is nevertheless important to note that by the 19th century, there was a convergence of thought in Europe that sought to proclaim the conquest of territories as a matter that was inevitable and God-given. Writing in 1853, even Karl Marx argued that 'England had a double mission in India: one destructive, the other regenerating – the annihilation of old Asiatic society, and the laying of the material foundation of Western society in Asia.'[3] When 'Asia' is replaced by 'Africa', the thought pattern is even more explicit. Indeed, there was the thinking that massive destruction of cultures and nature would ultimately be to the benefit of the plundered.

Conquest meant division and disunity across the continent. Conquest meant the splintering of nations and kingdoms into different blocs; it meant the amalgamation of disparate units into new wholes in a tensile state that promised no peace. Conquest and division laid prostrate vast civilisations in Africa, the Americas and Asia, siphoning off resources to fuel the industrial revolution in Europe.

Genocidal actions of one African ethnic nation against another cannot be separated from this history. In any situation where nations or groups compete for power or access to resources, conflict can rise from mere murmurs into open fratricidal wars. The goriest case of genocide in recent history is the one that occurred in Rwanda. Archbishop Tutu visited Rwanda a year after the genocide: 'I saw skulls that still had machetes and daggers embedded in them. I couldn't pray. I could only weep.'[4]

From the early years of colonialism, fissures on the continent separated Africans into boxes: anglophone for those under British rule, francophone for those ruled by France, lusophone for those under the thumb of the Portuguese. Of course, there were others who fell under the rule of the Italians, Belgians, Germans and Spanish. National boundaries were drawn arbitrarily, sometimes using a ruler rather than knowledge of ethnic national affinities on the ground. Not that that would have mattered much to the conquerors, whose sole aim was unfettered access to resources needed by the capitals in Europe.[5]

While Britain tended to favour the strategy of indirect rule in running her colonies, France preferred to go the way of assimilation. Through indirect rule, the British made use of local middlemen or compradors to run the territories while maintaining a grip on the central power structures. Such middlemen included local chiefs and a rising group of elites. The French, on the other hand, drew in the Africans in their colonies, giving them a sense that they were citizens of France when they were nothing of the sort. Assimilation included seats for African representation in the French parliament.

Over time, the stranglehold of the United States over Latin America as well as the strong influence of Russia and China over Asia were seen as strong reasons for a Eurafrica[6] union. In an article on the subject dated 1957, *Time* magazine commented on the Eurafrica proposal, referring to the statement by Christian Pineau, French foreign minister:

> Pineau said France has vast and beneficent plans not only for Algeria but for all its African territories. Said he: 'On the day when the [European] Common Market . . . has been created, [France] would like to promote the formation of a Eurafrican whole. Europe in its entirety, bringing to Africa its capital and its techniques, should enable the immense African continent to

become an essential factor in world politics.' Pineau's vision of Eurafrica did nothing to dampen the perfervid anti-colonialism of the Arab–Asian countries.

Leaders who pushed this idea included African statesmen such as Léopold Senghor of Senegal, who perceived that the relationship would be mutually beneficial. Detailed provisions were made in the Strasbourg Plan of 1952 but were rejected. To the most powerful Europeans, Africa was clearly an excellent backyard where they could exert continued influence, extract resources to meet their needs and also mobilise a ready army when combat situations warranted it. They were not, however, ready to give other European states unfettered access to the resources and trade in territories under their control nor to become involved in the colonial politics of other European governments.

Slaves and other energy sources

Access to raw materials and cheap labour made the plunder of Africa irresistible. One slaver was quoted as saying that slaves were 'free'; all you needed was to gather them in.[7] Bloodletting and easy dispensation of native lives meant nothing. Thus the early drive into Africa was fuelled by a liberty to do as one pleased within the sandwich of commerce and conquest.

The level of labour exploitation that occurred in those early years is unimaginable when considered today. Take as an example the exploitation of tin ore in the Plateau area of Nigeria. According to James Coleman, before a railway line was built to the mines in Jos, Nigeria, 23,000 Africans had to carry tonnes of the ore on their heads over a distance of 320km.[8] Africans also covered vast distances to fight wars that were not any concern of theirs: 374,000 Africans served in the British army during the Second World War and many were sent to the front in far-off Burma. It is on record that the bodies of fallen African soldiers were mutilated by some of Hitler's soldiers while at least 3,000 African prisoners of war were massacred in France. It would appear that the African soldiers not only fought other people's wars, but also became objects for sport and barbaric entertainment.[9]

There were other dislocating effects back home. In Sierra Leone, for example, the conscription of young men who had been engaged in farming, led to food shortages in 1919. A cup of rice that previously went for one penny was then sold at five pence. The impact on workers was particularly hard and they pressed for wage increases. The government imported rice to lessen the impact of the scarcity, but it was too little, too late.[10]

Africa offered ready sources of raw materials as well as a market for finished products. The stimulation of Africa's appetite for foreign goods was ensured by exertion of political control and imposition of Western cultural and consumerist norms. The strength that comes from unity, and the potential inherent in using one's own raw materials for development, were duly denied any conquered territory.

Colonisation of the continent can be understood as a stage in relations whereby the European powers took over the roles that had earlier been played by commercial entities such as the Royal Niger Company, which held sway in the Niger Delta. The power of America's Firestone Company conveniently ensured that Africans living in what is now known as Liberia were free from formal colonial rule. In southern Africa, the emergence of colonial authority relieved the British South Africa Company of direct political and administrative distractions. The companies stepped up their work of exploiting resources while the colonial governments provided security as well as the necessary backdrop for unhindered profiteering.

This model continued working into the so-called post-colonial (actually neo-colonial) era. In fact, the companies, aware of the rent-seeking nature of resource-financed and dependent regimes, exploit this vulnerability, while reaping massive profits. Things have never been better for companies, as their officials lord over the corridors of power, thereby ensuring that government policies further their interests irrespective of the impact on the people supposedly represented by these governments..

As the colonial era took its first steps in Africa, a handful of Europeans held the levers of power over millions of local peoples. After a few wars and a number of skirmishes, they had the continent prostrate before them. Thus it was that fewer than 2,500 white people who lived in Rhodesia had claim over 50 per cent of the entire land area before 1980, while 87 per cent of the land of South Africa was taken up by settlers.[11]

The overrunning of Africa did meet with resistance, but it was overpowered through sheer firepower or the subtle deceptions of the invaders and betrayals by compatriots. One thing cannot be denied. Spears can hardly withstand canons. While the spear or arrow may whistle silently, the canon boomed with a noise that could intimidate a strong warrior with no experience of such weapons of mass destruction.

Another factor that may have made the invasion easier is the intrinsic belief of the African that a person's humanity is inextricably linked to the humanity of the other person. This is the philosophical construct known as *ubuntu*. Archbishop Tutu captures this concept:

> The first law of our being is that we are set in a delicate network of interdependence with our fellow human beings and with the rest of God's creation. In Africa recognition of our interdependence is called *ubuntu* in Nguni languages, or *botho* in Sotho, which is difficult to translate into English. It is the essence of being human. It speaks of the fact that my humanity is caught up and inextricably bound up in yours. I am human because I belong. It speaks of wholeness; it speaks about compassion. A person with *ubuntu* is welcoming, hospitable, warm and generous, willing to share. Such people are open and available to others, do not feel threatened by others, willing to be vulnerable… They know that they are diminished when others are humiliated, diminished when others are oppressed, diminished when others are treated as if they were less than they are. The quality of *ubuntu* gives people resilience, enabling them to survive and emerge still human despite all efforts to dehumanise them.[12]

It is rather romantic to assume that Africans were inclusive in their relationship with the Europeans. The dehumanising of Africans must have had a debilitating effect on a population that trusted the good side of humanity. This can best be seen in the severe injuries inflicted by slavery. Africa's loss was the slave merchants' gain. It did not quite matter how many lives were lost in the voyages across the treacherous seas. All that mattered was that some of the cargo should make it to yonder shores.

And so it was that a company such as the Royal African Company could pay its shareholders 300 per cent dividends when out of 70,000 slaves they shipped between 1680 and 1688 only 46,000 made it alive. Many of those who did not make the crossing died of epidemics while others committed suicide by refusing to eat, or by hanging themselves with their chains or by jumping overboard to the embrace of circling sharks.[13] These suicides were not acts of cowardice. They were the epitome of the rejection of self-diminution.

The pathway of colonialism was not strewn with peace and easy pacification. But colonialism and capitalism both sought to extract the most resources with the least investment and costs. To a large extent the extraction of resources from conquered territories was and has been nothing but sheer robbery. Marx wrote in *Capital*:

> The discovery of gold and silver in America, the extirpation, enslavement and entombment in mines of the indigenous population of that continent, the beginnings of the conquest and plunder of India, and the conversion of Africa into a preserve for the commercial hunting of black skins, are all things which characterise the dawn of the era of capitalist production. These idyllic proceedings are the chief moments of primitive accumulation.[14]

The dependence on free or cheap labour fuelled the slave trade. The slave trade provided energy for the extraction of natural resources and for agricultural production. Slaves were the vital energy sources of capitalism. Africa was a prolific source of slaves and as these fuelled the machinery of foreign lands so the streets of Africa were plied with wailings and mourning. Great fortunes were thus built on genocide and on a massive exploitation of nature through what has been termed ecological imperialism.[15]

Taxation of colonial subjects and protectionism were ready tools that kept the colonies feeding conspicuous consumption in the metropolis. These also triggered resistance. Notable was the Salt March defiance championed by Mahatma Ghandi in India. In Eastern Nigeria, Aba women protested against unfair taxation and the destruction of their livelihoods in 1928–30. The women resented the economic hardship that taxation of their husbands and sons inflicted upon their families. While the men took to the bush on sighting the hated tax collectors, the women rose in defiance that became so heated that they went so far as to break open prison doors in Umuahia and force warrant chiefs to surrender their caps.[16] There are many such stories of heroic resistance across the continent.

Africa has always been a rich continent, even if the population does not share in or show any of that natural wealth. There is typically a disconnect between resources claimed by the state and the resources of the people, not even to mention the intrinsic value of nature, which can never be reduced to the functionality of resources. This disconnect arises from a number of factors, primary among which is the loss of people's sovereignty to the political and military apparatus of state. Citizens step out of the national economy, live autonomously as much as possible, and thereby make nonsense of GDP calculations and other indices by which development is measured in official circles.

The abrogation of the peoples' sovereignty began during the initial clash with foreign rule, where citizens had no say in who ran their affairs but simply had to submit to authority. This is the ember that smoulders beneath the hearth of despotism, where the proverbial strong men reign supreme. Leadership without accountability has always favoured reckless exploitation of resources and readily accedes to warped concepts of a free trade that is anything but free and is certainly not fair. Trade in African resources has largely been enforced by external forces that set the price of raw materials – which are exported – and, equally, fix the price of the finished products that return after processing abroad.

Patrick Bond has shown that unfair trade and investment relations are amplified by the liberalisation imposed by international institutions: 'trade liberalisation's damage is not limited to the primary product export drive with all its adverse implications. In addition, African elites have lifted protective tariffs excessively rapidly,' which destroyed local industries.[17]

The development of the field of 'economic development' itself is worth thinking about here. At the close of the Second World War, the debates on development at either the General Assembly or the Economic and Social Council (ECOSOC) of the United Nations were mainly about the concerns of industrialised nations. As discussions moved on to include the concerns of other nations, paternalism was the order of the day. At least one delegate reportedly stated his pleasure at his ability to bring to African people 'a mode of western reasoning'.[18] In such debates the interests of the colonial power were paramount, just as an authoritarian father would deal with his child irrespective of how the child felt. According to Pierre de Senarclens, 'The "civilisatory mission" of the West was extended due to efforts undertaken to develop countries under trust, capital investments and economic technical aid. In 1951, Mike Mansfield, the American delegate, placed this policy in the continuity of missionary activities at the Assembly.'[19] The dimensions of development that were conveniently ignored included distribution of power and control over natural resources.

As a result, the so-called resource curse can be traced not only to the corrupt, despicable dictators, whose spirited-abroad wealth often exceeded their countries' national external debt, but also to neo-colonial relations. Blaming a resource curse purely on dictators, as do some Western politicians, is a refusal to admit that the colonial pillage of Africa continues, driven on the same tracks that were set in those dark days by transnational corporations, trade rules, bilateral and multilateral arrangements, powerful international agencies such as the World Bank and the International Monetary Fund (IMF). These forces retain a vice-grip on Africa, impose structural adjustment programmes that stifle development and then drop a few coins in the form of aid into the hands of a devastated continent.

Over the years the story has remained the same. Communities and nations are dislocated from their material means of production, separated from their systems of sustainable livelihood and made to become bystanders on the dusty byways, as rulers and international agencies spout about development and progress. While nations struggle to meet local food needs, more chains are introduced through expert advice that encourage them to use arable land for agricultural production for the benefit not of local citizens, but for export. After centuries of plunder, what is there left in Africa to attract adventurers and seekers of El Dorado?

2

Africa is rich

Sacred land
Your defenders
Link hands across generation gaps
Kids, ancestors, butterflies merge confronting the rage of crude
addicts
Impotent capital halted by guardians of your treasured space
Still you stand still
Your calm visage shocks me
Your verdant mane, rivers of life
Mother Earth's best patch besieged[1]

AFTER UNINTERRUPTED PLUNDER, Africa still remains rich. If the slavers of old had to cut through uncharted territories to reach the cargo, those who followed on the colonial train had to devise means of co-opting accomplices on the continent. Over the years the continent was charted for its usefulness, with railways and highways and ports facilitating extraction and packaging for waiting consumers and manufacturing plants in the North. This story remains the same to this day.

Walter Rodney's masterpiece *How Europe Underdeveloped Africa*[2] argued that although capitalism could revolutionalise agriculture in Europe it did not do so in Africa. Colonialism was incapable of improving agricultural tools for the same reason slave labour was not appropriate for the factory floor. The international division of labour requires skilled labour in the metropolis and low-level manpower in the dependent colonies. Rodney concluded, 'The most convincing evidence as to the superficiality of the talk about

13

colonialism having "modernised" Africa is the fact that the vast majority of Africans went into colonialism with a hoe and came out with a hoe.'

The reality is that today Africa is little more than a mere provider of raw materials. This reality is built on the global structure of relationships deeply rooted in inequality, for as Eduardo Galeano observed, the United States expanded in territorial size and was highly protectionist in trade relations, while the colonies in Latin America were fragmented and also forced to liberalise trade, opening up their territories to cheap imports that emasculated nascent manufacturing. Nevertheless, the 1930s–40s were a moment when economic and political crises in the North permitted partial delinking, balancing and growth in parts of the South – even settler colonies like South Africa and Zimbabwe – leading to the 1970s school of dependency theory, which advocated inward-oriented growth.

Since the 1980s, protectionism has remained the preserve of the North and every tool is used to preserve that privilege and to force Southern nations to create so-called free trade zones as well as enter into a diversity of free trade arrangements. Some of these arrangements, promoted by international financial institutions, include the infamous structural adjustment programmes (SAPs) that decimated African economies in the 1980s and 1990s. They include instruments such as those extended to highly indebted poor countries, known as poverty reduction strategy papers. The World Trade Organisation and its rules further illustrate how blatantly unjust the global trading regime is.

SAPs led to privatisation of state-owned enterprises under the guise that only the private sector can operate in an efficient way. The implication of this doctrine is that the public sector incubates, births and hands over institutions to private hands well connected to the system. As happened during the financial crisis that hit the world in 2008, when private enterprises were mismanaged, it takes public funds to feed and nurse them back to life and hand them back into the sleazy hands of the same folks who ran them aground in the first place. This is like handing back an abused child to the perpetual custody of an unrepentant, abusive step-parent.

Another manifestation of the privatisation strategy is that publicly owned enterprises are sold to transnational companies either directly or through local fronts. This stands out as an easy way to snuff out opposition where both the buyer and the company on the auction block are in competition. When transnational corporations grab local companies they also abort the possibility of self-determined economic development.

Yet another consequence is that the SAPs insist upon export promotion under ridiculously slanted conditions that are detrimental to the implementing country. They generally make countries lose control of their economy through removal of trade barriers, currency devaluation and retrenchments in the public service. They also enforce cuts in spending on social services. These inexorably move countries further into the debt trap.

The debt trap has been another effective means of keeping the raw materials marching northward. As African nations groan under the debt burden, the servicing of these debts requires more and more foreign exchange. While the debt remains unpaid, nations have to accept unrestrained resource extraction so as to satisfy the unyielding demands of the creditors. Efforts to keep up with the demands and to free debtor nations can superficially be compared to a dog chasing its own tail – but it would be more apt to compare this to a new form of slavery, where the slave actually believes that he is free.

Most violent conflicts in Africa can be traced to the scramble for her natural resources. These conflicts manifest themselves in different ways and with varying intensity. There are also several underreported cases where irresponsible extractive industry operators rip through the continent, grabbing what they can and leaving the land scarred and the people impoverished. Consider the conflicts that brought Liberia and Sierra Leone to their knees; the conflicts that ravage Central Africa's Great Lakes region and the Sudan; and war in the oilfields of Nigeria's Niger Delta.[3] We start with the tragic case of the Democratic Republic of Congo (DRC).

Congolese chaos

Writing on the Congolese war of the 1960s, Che Guevara noted, 'Katanga's mineral wealth made it the key area of the Congo and the one where the toughest battles had to be fought.'[4] These conflicts are real and direct drivers of power relations

and continue to smoulder because of the great profits that some
merchants derive from the ensuing human misery.

Reviewing how Western imperialism was fuelling conflict in
DRC in late 2008, Matt Swagler noted,

> The latest fighting in eastern Democratic Republic of Congo
> (DRC) isn't the result of ethnic rivalries, as portrayed in the
> mainstream media, but the logical outcome of intervention
> by Western governments and profit-seeking corporations. The
> U.S. is fuelling both sides of the conflict – by backing neigh-
> bouring Rwanda's support for rebel forces on the one hand,
> and a United Nations 'peacekeeping' operation in support
> of national DRC troops on the other. Clearly, peace for the
> Congolese people is second to securing U.S. economic and
> political interests in the region.[5]

The Congolese resource wars have sometimes been characterised
as a civil war, of rebel groups fighting because of ethnic cleavages, but the fact is that the conflicts have been orchestrated
by greed for her rich resources. The smouldering conflict could
ignite a major conflagration echoing past conflicts in which over
5.2 million Congolese lost their lives between 1998 and 2004. That
horrendous war was, as Leo Zeilig has remarked, 'the bloodiest
conflict since the end of the Second World War'.[6]

The wealth of the Congo basin is of legendary proportions. The
area has some of the largest remaining rainforests on the continent and vast quantities of solid minerals waiting to be extracted.
The DRC's Kivu Province is loaded with gold, coltan, cassiterite
and others. Rwanda, Uganda and their supporters in the United
States have deep interests in keeping the extraction corridors of
eastern DRC open for business in a thick cloud of gunpowder and
a swamp of blood.

Large-scale violence and plunder in the Congo characterised
Belgium's colonial reign under King Leopold. The agents of this
Belgian king, who never visited Africa, enslaved Africans and
used them to harvest ivory and tap rubber. Those who did not
meet his expectations had their hands hacked off. This pattern
continued, and today nations and corporations hungry for the
mineral resources of DRC are suspected of funding militias and

keeping the fires of conflict burning in the region. The continent's largest corporation, Anglo American, was caught red-handed paying off eastern DRC warlords and its chief executive, Bobby Godsell, merely remarked, 'Mistakes will be made'[7] – leaving the impression that the mistake was getting caught.

It sounds incredible, but as Swagler recounts:

> In 1996, the U.S. sponsored an invasion of the DRC from Uganda and Rwanda based on satellite maps provided by San Francisco's Bechtel Corp. Using these maps, an estimated 800,000 Hutu refugees were hunted down and killed in the eastern forest regions. Some of these refugees died directly at the hands of U.S. Special Forces and private mercenaries from the Virginia-based Military Professional Resources Inc., housed at a U.S. military base in Uganda ... In 2006, Keith Harmon Snow reported that $6 million worth of cobalt (used largely in jet engine production) was exiting the country each day, mostly headed for North American and European bank accounts. The yearly value of this cobalt extraction alone is greater than the DRC's entire federal budget.[8]

The *Wall Street Journal's* Robert Block stated that Bechtel's generous assistance facilitated 'the most complete mineralogical and geographical data of the former Zaire ever assembled, information worth a fortune to any prospective mining or oil firm'.[9] Indeed, the company is as renowned for engineering and construction as it is for providing intelligence services and data to the US government. Revolving door personnel includes George Schultz (former Bechtel president and Reagan's secretary of state), Richard Helms (former CIA director and later Bechtel consultant) and Steve Bechtel (CIA liaison to the Business Council).

A report titled 'The business of war in the Democratic Republic of Congo: who benefits?'[10] unpacked how Western businessmen flocked to Kabila during what was known as 'Africa's First World War'. According to World Policy Institute researcher Dena Montague:

> The International Monetary Fund (IMF) and World Bank have knowingly contributed to the war effort. The international lending institutions praised both Rwanda and Uganda for

increasing their gross domestic product (GDP), which resulted from the illegal mining of DRC resources. Although the IMF and World Bank were aware that the rise in GDP coincided with the DRC war, and that it was derived from exports of natural resources that neither country normally produced, they nonetheless touted both nations as economic success stories.[11]

Block disclosed that one executive from all others obtained primacy of position as Kabila's adviser and travelling companion: Bechtel's Robert Stewart. None of this is averse to Bechtel's own corporate anthem. As Steve Bechtel said, 'We are not in the construction and engineering business. We are in the business of making money.'[12] What better place, corporations may ask, than the DRC, described by Africa's colonial 'discoverers', once upon a time, as the 'geological scandal'.

West African riches

The militias active in the DRC do not shrink from enlisting children into their ranks as soldiers and as slaves. A similar pattern was evident in Sierra Leone's bloody civil war. The lure of diamonds, that glittering carbon, caused men to enlist kids into militias, arming them with AK47s and machetes, cutting limbs as though they were slashing through sugarcane stands. I visited recently, to see what it meant to have peace under conditions of mineral exploitation, and on the way travelled through Mali and Guinea.

The road to Morila in the region of Sikaso, Mali, is not one you would doze off on. I did that seven-hour journey with friends on the West African Advisory Board of the Global Greengrants Fund. Our lead man was Dembele Souleymane of Friends of the Earth Mali. We had been meeting in Bamako and decided to head south for a feel of what was going on in the mining communities. A couple of years earlier we had been in Bamako and had taken a trip to the Tombola community on the border with Guinea. On that occasion I took the liberty of crossing a nondescript bridge over a little river and marched into Guinea without border controls of any kind. In Tombola almost everyone had the surname Kamara.

The route to Tombola was memorable for the picturesque rock formations along the highway. Additional interest was getting to

2 AFRICA IS RICH

a stop along the road where the famous king of the Mali empire, Sundiata, is said to have visited. The mining in the Tombola axis was mainly artisanal, with locals punching holes in a wide expanse of land, seeking the grains of gold that brought smiles to many. It was an upbeat community where we enjoyed curried rice in wide basins and for dinner had chickens that had been presented live to us an hour or so earlier. As we reflected on the mining going on there, we asked why there were so many young folks in the pits rather than in school. It turned out the young boys and girls dreamed that they would strike lucky and make enough to either pay their way through basic education or pay a decent dowry when the time was ripe. Hope.

This hope drove the community to create dangerous pits that were abandoned over a wide area, posing great danger to both man and beast. Each pit was about 90cm in diameter and up to 800cm deep – yawning pits of destruction with neither warning signs nor barriers.

Before moving on to Morila, we had meetings with representatives from about 40 neighbouring communities and shared experiences with them about the challenges of living in the shadows of wealth. One challenge here was a leakage of cyanide into the landscape and probably into one of the water bodies on which some of the communities depend for potable water. We got to Morila at night and were told we could check into the mayor's guesthouse, but first we were to pay a courtesy visit to the mayor. The mayor was an affable fellow in a humble brick home with a corrugated metal roof. We were not welcomed in that house. We were received in a more traditional open-sided thatch house into which you literally had to bow low to be able to enter. Perhaps a way of paying obeisance to the mayor?

After pleasantries were exchanged, we sought to learn whether any fraction of revenues accruing from the gold mines made their way to the communities. It turned out that a fair chunk was controlled by the mayor's office, and used mainly for administration of the district. There was no electricity in the village and the mayor had to make use of a private electricity generator. The mining company extended a light point to the village square where the mayor's office and the guesthouse were located. When we got to the square it was already dark except for the solo light pole extended there from the miner's camp. Under the floodlight, some kids were playing soccer with so much gusto it was immediately easy to see why Mali is such a good soccer nation.

Morila is reputed to be the second largest gold mining community in Mali after Sadiola. This is the base of RandGold, a company that was said to have experienced the most frequent clashes between mining company workers and management, and on one occasion between the company and local population. In 2005, the company fired more than 200 workers for their alleged role in 'inciting' workers over a pay rise.

The village that owns the land where the gold mine sits is now in the shadow of the rising pile of residues from the gold ore, known as tailings. It is a small collection of a few homes and a health centre that locals claim caters mainly to junior workers from the mine. The visible sign that this village is actually near a mine is the heavy coat of dust that rests on it. When Rand-Gold first arrived the people were hopeful that they would gain employment with the company. At the outset some villagers were hired as casual labour to clear the bush so that work could start on the concession. However, soon after the initial bush clearing the gates were shut against the locals.

When we asked the mayor why they did not demand better environmental standards and economic rewards from the gold company, the answer we got was telling: who can fight a huge mining company? But some are doing just that as we will see.

The groundswell for resistance in Kono is somewhat tentative, as can be perceived in Morila or Tombola. Kono, a war-ravaged town in Kono district, in eastern Sierra Leone, is known for its diamond mines. Indeed, in these parts almost everyone dreams of stumbling on a piece of the rock. During the Sierra Leonian civil war, Kono was mostly held by the rebels and was severely damaged. When I visited Kono on 10 November 2009, the anniversary of the murder of Ken Saro-Wiwa of Nigeria, I could not help wondering when non-violent resistance would rise here. Or would it?

The drive from Freetown to Kono was smooth until we got to a point where the road ceased to exist in the usual sense. The stretch from there to Kono would ordinarily be covered in, say, 90 minutes, but it now takes four hours if you are adept at driving through craters. The traffic is tellingly light, apart from some four-wheel-drive vehicles and the occasional car meant for five people, but generally loaded with ten or more.

The vegetation is lush, the hills beautiful. No wonder many say that Kono district was the breadbasket of Sierra Leone. If it still is, then it must be tough getting the bread or the basket to other regions where the produce is needed. Severely rattled, jolted, jostled and knocked about in every direction, I decided there must be something I could do to distract my attention from the heinous road. Soon I began to take note of the wildlife that crossed the road or scurried along. Within a short space of time I counted four fat squirrels hopping to only squirrels know where. Within the same space of time I saw three guineafowl and five crows. Not long thereafter we happened on a flock of crows and I knew counting them would be a hopeless task. On the bumpy ride back I started to note the wildlife again. We saw squirrels, an armadillo, crows, a giant rat and then a flock of guineafowl marching majestically across the road.

Things changed for Kono in 1930, when the first diamond was discovered at Gbobora stream. Since then diamond activities have dominated the socio-economic life and livelihood of people in the district

The scars of war are visible as many houses remain without roofs. Land degradation, water pollution and other environmental concerns allegedly provide talk shops for bureaucrats and politicians. Kono, once a food basket, is now characterised by diminished agricultural production as many of the citizens have abandoned the farms for mines. Consequently, there are quantities of uncovered mining pits, degraded diamond fields, some of which are ten times larger than a standard football pitch, and ponds of contaminated water that are often breeding places for mosquitoes. According to a report: 'From Sewafe town to the heart of Koidu town and beyond, there is ominous evidence of environmental decrepitude. The mining is both alluvial and Kimberlite, mechanised and non-mechanised, and is evident in front of dwelling houses, back of houses, in gardens, around farms, paths, in streams, rivers, hills and valleys, and along motorways.'[13] Yengema town was once described as the most beautiful in the Kono district. Rather than have its beauty enhanced by diamonds, the landscape has taken a severe beating and it is now a shadow of its former self.

As incredible as it may sound, the miners have pulled down bridges and houses, and although most residents have expressed concern, they are afraid to voice them aloud because top politicians are behind these destructive extractive activities. The uncontrolled mining in Yengema town is also said to been condoned by the mines warden and the mines monitoring officer in Yengema, who have allegedly refused to bring the perpetrators to book. A member of Friends of the Earth Sierra Leone who accompanied me on the trip to Koidu, Diana T. G. Kamara, said that when she was growing up in the town, some miners could come to a house and tell the owner they suspected there were nuggets of diamond beneath their sitting room. After reaching an agreement, they would proceed to excavate and would indeed sometimes find diamonds. Whether legend or truth, I had no need to verify; it showed that township mining meant miners could knock people away from their dining tables, dig beneath their beds and uproot graves which happened to lie in an area suspected to harbour an errant stone, more precious than bones.

The Yengema Descendants' Union (YENDA) condemned reckless mining in the town. The group argues that the work is illegal, defying laws that stipulate that no mining activity can take place in residential areas. Houses in Yenkee Street in Yengema have been targeted and a mosque, which was rehabilitated through the support of the UN peace-building fund, is also under threat of collapse because of the diamond mining.[14]

Mining companies do not present friendly faces. Attempts made to view mining activities in the concession areas of the Koidu Holdings were rebuffed at the gate as the security personnel insisted that only the company top shots could allow access. Understandably so, as they may have chunks of diamond under their tables. But do they ferry the gems to the ports over the horrendous road network that links Kono to everywhere else?

My friend Michael Aruna, a resident of Kono welcomed us and together we visited some artisanal mining points in the city. The miners we met were all optimistic that perseverance pays. None of them said they had found a meaningful gem in the past week yet none were prepared to quit. One said he had abandoned his lucrative carpentry business because of an arm injury. Today he was panning for diamonds, using that injured arm dexterously. One would not have guessed he had any injuries.

Aruna tells tales of how mining has captured the psyche of the people in the town. He told me stories of resistance in the area and how Koidu Holdings marked several houses for destruction in order to mine for diamonds beneath them. The affected communities are to be relocated but little has happened yet. There is the story of one community activist who was offered 10mn leones (about $2,670) to relocate. The man rejected the money and insisted on being given a house, just like the one he had built for himself. After some months he was given a single-bedroom house, clearly inadequate for his large family and not comparable to the house that was being demolished. When he rejected this offer he was run out of town and now squats in Freetown as a mining refugee.

Koidu Holdings is a South African mining company set up by its parent company BSG Resources Limited for the purpose of exploiting diamonds in Sierra Leone. They are operating the Kimberlite project in Kono area under a 25-year lease agreement.[15] The Network Movement for Justice and Development (NMJD) foresaw the antics of the Koidu Kimberlite Project right from the start. Working under the general ambit of the Extractives Industries Transparency Initiative, NMJD demanded to know whether the Koidu Kimberlite Project run by Koidu Holdings was above the law. In a statement they complained that whereas the company sent copies of the environmental impact assessment it conducted to the Word Bank in October 2003, the same documents were only released in Sierra Leone in late January 2004. They asked pointedly whether the World Bank was now the key stakeholder in the area, displacing the people of Sierra Leone. In fact, NMJD claimed that even when the documents were said to be on display they were actually nowhere to be found. Here is an interesting extract from the press statement:

On Tuesday 3rd February, 2004 Peace Diamond Alliance (PDA) convened a meeting of its executive attended by the Deputy Minister of Mines, Deputy Director of Mines, the SDO-Kono, Parliamentarians, chiefs and the personnel of Koidu Holdings

Limited (KHL) in order to resolve the issues and concerns so raised. Inspite of the public outcry and effort of PDA, nothing has been done to address the problems. Instead, His Excellency The President went to Kono a few days after the PDA organised meeting praising KHL [for] 'a job well done' and urged the people of Kono to cooperate with Koidu Holdings Limited. This was followed by a television show displaying the more than 10,000 carats of diamonds already gotten by KHL in the midst of cries, pains and sufferings of the local people. Not withstanding this the Campaign for Just Mining and civil society of this country will not be deterred as they have the right to perform their constitutional duties, namely, to 'make positive and useful contributions to the advancement, Progress and well-being of the community' section 3(f) 1991 constitution.[16]

In response to the statement, Koidu Holdings claimed that they had done no wrong as far as the environmental impact assessment process was concerned. At the same time they questioned NMJD's right to speak on behalf of the affected communities.[17] Just as the company had first sent their environmental impact assessment reports to the World Bank, so the response to the NMJD press statement and campaigns were sent not to the NGOs, but to the Multilateral Investment Guarantee Agency of the World Bank. The strong links between mining corporations and international financial institutions cannot be denied.

It is not clear how much of the profit of Koidu Holdings – incorporated in September 2003, and owned by BSG Resources Limited through its subsidiary BSGR Diamonds Limited[18] – goes to the Sierra Leonean state. When commercial mining of diamonds began, the monopolies for iron ore and diamonds were held by two foreign companies, Sierra Leone Development Corporation (DELCO) and Sierra Leone Selection Trust (SLST), and they were obliged to pay the government a mere 5 per cent of whatever profits they made, if any. The companies also held leases that were for 90 years.

The rise of minerals as a major income earner in Sierra Leone follows the same trajectory as seen in other nations in Africa. By 1931 minerals made up less than 5 per cent of domestic exports. This rose dramatically to 43 per cent in 1950 and 86.7 per cent in 1961. At that time diamonds accounted for 60 per cent of mineral exports and 43 per cent of all exports.[19] While mining took the prime position, agriculture suffered a corresponding decline. A major staple in the country has been rice, and before mineral extraction pushed farmers off farms, the country was self-sufficient in rice. Rice shortages contributed to widespread riots in 1955–6, precipitated by a diamond rush in which hoes were abandoned for pans.

The webs of resistance keep growing. They may appear tenuous, but these webs are as resilient as those spun by spiders. The strength of these webs is created and reinforced by community memories of innocence lost. Having one's environment unrecognisably altered is an impetus for questions to be raised.

While our core values and ethical fabric are one and the same throughout humanity, we recognise that national values and a sense of ethics have been assaulted at personal, community, state and national levels, making it a mere commonsense conclusion that the matter requires a radical and holistic interrogation. As leaders appear before courts of law and as truth commissions and probes of different sorts take place in various parts of Africa, we see to what depth the sense of right has been dislocated. It becomes apparent that unless radical surgery is carried out we would be attempting to treat a terminal cancer with mere analgesics.

The conversion of public goods into private property through the privatisation of our otherwise commonly held natural environment is one way neoliberal institutions remove the fragile threads that hold African nations together. Politics today has been reduced to a lucrative venture where one looks out mainly for returns on investment rather than on what one can contribute to rebuild highly degraded environments, communities and a nation. This is one of the benefits that structural adjustment programmes inflicted on the continent – the enthronement of corruption.

The curses of minerals and petroleum

After centuries of plunder the mineral wealth of the continent remains massive – and thus the magnetic pull she has for fortune seekers. In the bowels of the continent are 75 per cent of the world's platinum group of metals (cobalt, chromium, etc), 50 per cent of the gold, 45–50 per cent of the diamonds, 25–30 per cent of the bauxite, 10 per cent of the nickel and copper, 12 per cent of the uranium, 7 per cent of the manganese and 5.8 per cent of tantalum (coltan). Coltan is found in the DRC, Nigeria, Ethiopia and Zambia.[20]

By the end of 2006 Ghana was among the top ten producers of gold in the world with 69.8 tonnes, Zambia was in the same bracket for producers of copper with 516 tonnes of the metal. South Africa was among the top ten in the production of both nickel and aluminium with 41.6 tonnes and 887 tonnes respectively. South Africa was the biggest producer of platinum (164.5 tonnes) and the second largest producer of palladium (90.4 tonnes).[21]

Africa boasts of probably no more than ten per cent of the world's proven oil reserves, yet she is the object of desire of oil-guzzling nations of the North, notably the United States and, increasingly, China. More rigs are clawing into the continent and more oil tankers and floating production, storage and offloading (FPSO) facilities are springing up in the area. For example there is the 114,000 tonnes FPSO scheduled to go into active service in 2012 in Elf's Usan field in waters almost 1km deep and 100km off the coast to the south east of Bonny Island, Nigeria. This FPSO is designed to store two million barrels of crude oil with the capacity to produce 160,000 barrels of oil daily as well as five million cubic metres of gas per day.[22]

With prodigious finds virtually all over Africa, investors are going overboard to wrap up deals. A first was recorded in 2009 off the coast of Congo when the world's first floating, drilling, production, storage and offloading (FDPSO) vessel was commissioned. The FDPSO is unique in the sense that it adds something the regular FPSOs do not have: it drills production wells. The drilling unit is designed in such a way that it can be detached when the job is finished in a particular location and moved somewhere else. Moored at a depth of 1400m, this FDPSO has the capacity to store 1.4 million barrels of oil, process 60,000 barrels of fuel and 40,000 barrels of oil per day.[23]

John Ghazvinian in his book *Untapped – The Scramble for Africa's Oil*, gives some reasons[24] why Africa's crude oil is so attractive. It offers a more favourable geographic location than other extraction sites – more of the new discoveries are found in deep or even deeper waters offshore and away from questioning communities. Another source of the allure is the fact that much of the crude oil found on the continent is of the variety tagged sweet or light. This is easier to refine than the so-called heavy crude. The crude is called sweet because it has a low sulphur content and is viscous. Refining it is easy and the profit is higher than for heavy crude. Although the crude may be called sweet, the effects on the local communities where the crude is extracted cannot be characterised as sweet.

Ghazvinian gives two other reasons for Africa's resource curse: one is the contractual environment that ensures oil companies can make a killing in their profits. Many countries in Africa accept production-sharing arrangements whereby oil companies invest in the exploration and extraction activities, recover their costs and then share profits with the countries. It is an arrangement through which impoverished nations can access a portion of the oily wealth beneath their feet. Within this contractual environment, the oil corporations declare how much oil is being extracted as well as what constitutes the production costs. Such production costs often include the toothpicks that oil company workers use in their cafeterias.

Sudan alone, before its split in July 2011, provided China with about 10 per cent of China's crude oil needs. Thus, while the nation burned and atrocities were recorded in Darfur, Abyei and other areas, no resolution at the Security Council of the United Nations would be raised without China threatening to use her veto power. In this relationship, oil is thicker than blood.

New oil states are emerging all along the Rift Valley of eastern Africa. As I write, Uganda is gripped by oil fever and hopes are raised high – as is often the case in new oil territories – but the higher hope is raised, the louder the thud when it crumbles. As more and more finds are made, formerly ignored countries are now receiving focused attention.

Take, for example, São Tomé and Príncipe, an island state and the second smallest nation in Africa. The population of this country stands at less than 200,000 today. Human habitation is reckoned to have started in 1470 with the arrival of Portuguese travellers who came with Africans from the mainland. The Portuguese ruled over São Tomé and Príncipe until 1975 when they eventually bowed out due to popular pressure and resistance from the citizenry when the Portuguese dictatorship was overthrown and other Portuguese colonies were liberated. Educational and other infrastructure remained meagre through years of political instability.

Before oil, the country depended largely on agriculture, including fishing, and on foreign aid. Ghazvinian describes the political culture of the island as 'handshake politics' and cronyism.[25] The country has experienced 14 changes of government since 1991 and it is hoped that oil production will not hasten the speed of the merry-go-round. Although serious oil exploration activities began in the 1970s it was not until 1995 that it was announced that ExxonMobil had made a big find in the Zafiro field between the country and Equatorial Guinea. A Texas-based environmental remediation company, Environmental Remediation Holding Company (ERHC) entered on the scene in 1997, offering the government the sum of $5mn for the right to act as middleman in oil-related licence negotiations on behalf of the government. ERHC metamorphosed into an oil exploration company in 1996 although investigations revealed that ERHC had only one paid member of staff, no drilling equipment and only $1.5mn in cash.

When ERHC ran into stormy waters with the government the resolution of the conflict led to its controlling shares being taken up by Chrome Energy, a Nigerian company with deep political connections and influence on both shores. ERHC/Chrome Energy arrangements with the São Tomé and Príncipe government in 2001 allowed the company many privileges, including a 15 per cent stake in up to four oil blocks as well as 10 per cent of whatever profit the country might make from oil in the future. This neo-colonial arrangement also made the resolution of the maritime boundary dispute with Nigeria a key conditionality for the takeover, generating the Joint Development Zone between the two countries as a prime spot for cornering petrodollars.

And what of Ghana? In July 2009, President Obama made the point of visiting Ghana on his first trip to Africa after being elected as president of the United States. Although the choice was ostensibly predicated on Ghana's democratic credentials, having recently conducted a closely contested presidential election in which the opposition won, some analysts do not dismiss the possibility that once again Africa's pull was the new oil discoveries, both onshore and offshore. Obama's visit appeared to be the landing of an emphatic boot on African soil to indicate how vital the Gulf of Guinea is to Washington's energy security. With the establishment of the African Command in October 2008, no one is left in doubt that the US means to keep energy supply channels open by any means necessary.

Ghana's offshore oil find, the largest in the past decade in the region, is named Jubilee Fields. The dream of an influx of petrodollars is already pulling a train of business speculators into the country. But Jubilee Fields, predicted to hold 1.2 billion barrels of oil, is firmly in Texas-based Kosmos Energy's kitty. As Emira Woods of the Institute for Policy Studies explained:

> In May, 2009 Kosmos began to draw bids for shares of its stake in the oil-rich fields. Global energy players – Chevron Corp, Exxon Mobil, Royal Dutch Shell, China National Offshore Oil Company and British Petroleum – all with a focused eye on Africa, and a bloodied record on the continent, are beginning to circle like vultures. After all, the deadline for Kosmos Energy Bids is July 17, a week after Obama's visit to Ghana.[26]

And as appears usual in Africa, opacity and secrecy jurisdictions (also known as tax havens) remained central to the greasy wheeling and dealing surrounding Africa's resources.

The *London Review of Books* claimed in a short blog piece:

> The *Kwame Nkrumah* MV 21, the Floating Production Storage and Offloading facility that will be used to exploit Ghana's offshore oil during the first phase of development, is owned by Jubilee Ghana MV 21 BV, a special purpose company incorporated in the Netherlands. The Netherlands is host to more than 20,000 'mailbox companies' (of which 43 per cent have a 'parent' in secrecy jurisdictions such as the Cayman Islands, the British Virgin Islands, the Netherlands Antilles or Cyprus). It specialises as a 'passthrough' conduit for financial flows including 'dividends, royalties and interest payments' via 'special financial institutions'. The Dutch Central Bank defines ring-fenced SFIs as 'subsidiaries of foreign parent companies used to channel capital through our country, which has little influence

> on the Dutch economy.' The Netherlands does not place details
> of trusts on public record, or require that company accounts or
> beneficial ownership be made available for public record.[27]

The story of recent privatisations in the mining sector shed more
light on how Africa's resources continue to be exploited. As the
SAPs of the World Bank and the IMF swept across the continent
in the 1980s and 1990s, Ghana's leadership was quick to liberalise
the economy. Such policies had a devastating impact on agriculture. The mining sector witnessed increased productivity but no
value was added to the communities in whose soil the activities
were undertaken. We will look at the pioneer commercial mining
company in Ghana, Ashanti Goldfields Company (AGC) – now
AngloGold Ashanti – and how the reforms affected its operations.

The history of AGC began in 1897 and since then it has played
a key role in the crucial gold-mining sector. From the 1970s to the
early 1980s, gold accounted for 80 per cent of the mining sector's
export earnings. By 1994 the gold sub-sector brought to the country a hefty 45 per cent of its entire export revenue, well ahead
of cocoa, which brought in 25 per cent of total export revenue.
Earnings from gold rose to 94.5 per cent of the total revenue from
mineral exports in 1997, with diamonds, bauxite and manganese
bringing in the balance.[28]

The privatisation of the Ghanaian mining sector was preceded
by a sector-specific programme including a Minerals and Mining
Law in 1986 'to provide the conditions demanded by multinational mining companies for their participation in the country's
mining sector'.[29] The law, which has since seen further revisions,
provided the industry with generous incentives and reduced
the risks to be borne by the investors. As part of the reform
programme, government shares in AGC were reduced from 55
per cent to 19 per cent in 1998. As the sector became deregu-
lated, more foreign direct investment flowed in, amounting to
$4bn between 1983 and 1998. The result has included a fourfold
increase in output from the sector between 1990 and 2002.

Although mining brings in a good chunk of the export earnings, the sector does not have a commensurate link with the
domestic economy since, by its nature, it operates as an enclave
industry, with generous capital allowances. To underscore the

minimal impact of the mining sector on the domestic economy it is noted that mining has contributed a mere 2 to 5 per cent to the national GDP since independence, while contributions from the agricultural sector reach 36 per cent. Employment in mining represents 5 per cent of formal employment in the country and although wages in the sector may appear to be higher than in other sectors, this is primarily a result of the high wages earned by expatriate staff.[30]

At one time, AGC was the sole substantive transnational corporation from sub-Saharan Africa, with tentacles in Burkina Faso, Côte d'Ivoire, DRC, Ethiopia, Guinea, Mali, Senegal, Tanzania and Zimbabwe. After Anglo American purchased AshantiGold, its base shifted to Johannesburg, South Africa, since the South African authorities liberalised investment relations in 1999 as part of the county's home-grown SAP. It is noteworthy that from the introduction of SAPs into Africa, at least 40 African countries have changed their resource sector investment codes in order to attract foreign investment.[31] This rapid compliance was largely achieved by making investment code revision a condition for the granting of loans.

We cannot leave the gold bars of Ghana without a look at Obuasi, a town that has played host to AGC for over 100 years. T.M. Akabzaa describes communities where open-cast mining takes place as sites with thick clouds of dust hanging in the air. But Obuasi's mines are primarily hidden beneath the town and the outward, telltale signs are the mountains of tailings in the landscape. Akabzaa describes Obuasi thus:

> The shafts and the dunes of mine wastes are indeed a rough measure of the amount of wealth that has been extracted from beneath the area in the last 107 years ... Indeed, the entire township of Obuasi rests on a slab, which doubles as a roof to these huge underground openings. These openings accommodate scores of gigantic dozers, loaders, compressors and dumper trucks ... hard to miss is the setting of the modern accommodation facilities for management and middle-level staff of the mining company at the summits of some of the hills. These beautiful buildings contrast sharply with the shanty structures within the low-lying areas of the township, which host the peripheral businesses. Another noticeable feature of the

31

township is the contrast in its road infrastructure. All asphalted roads lead to AGC's facilities such as the company's hospital, senior staff residences or the commercial centre hosting banks, the post office and prison service within the fringes of the mine. The dusty and pot-hole riddled roads in the township on the other hand, lead to public sector facilities including the public hospital, the office of the district assembly, private residential areas and local satellite communities.[32]

These words were written in 2007. When I visited Obuasi and the outlying villages in October 2009 with Abdulai Darimani of Third World Network-Africa and other friends, the situation had degenerated. And that is not surprising. Mining towns in the South generally experience some boom seasons and then are left gasping for breath and eventually become a shadow of the promise they once held out to their people.

At the Binsere village area the pains of the community were palpable. A huge open-cast mine was abandoned at this community. According to a community representative, the pit was as deep as 100m at a point. A nearby school building had to be abandoned by the community because the structure became unsafe due to blasting for gold. At the time of my visit the pit was about half that depth, having been filled with waste laden with cyanide from other mines. In other words, AngloGold Ashanti simply turned this area into a waste dump and, after partially filling it with more waste, declared that the environment had been reclaimed.

Dangerous as this dump is, some artisanal miners were busy digging up the toxic mud in search of gold. Perhaps they had believed AngloGold Ashanti's claim to have reclaimed the environment and thought all was safe? When asked if they made a living from such a deadly venture, the miners said that they were simply gamblers, who had no choice. They could sift through the mud for a week without an ounce of gold, and then suddenly hit a few grams that would put a little food on their table. Did they know the toxic mud was killing them by instalment? They did. This yawning pit of death tailed off into a lake about a kilometre from where we stood on the shoulders of the mine. Being told that this situation was better than that at the next village, I was eager to get there.

Apart from the spatial disparity in the development of the mining community, the health impacts of the mines present another unacceptable price to pay for some shiny metal on one's fingers or tooth. The tailings and waste ponds from Binsere and other mines in the Obuasi area pass their toxic cocktail into the stream at Dokyiwaa community. This environment – both land and water – is so polluted that the community had to be relocated.

Research by TWN-Africa showed that both the ground and surface water in the Dokyiwaa area have been severely contaminated and this has equally affected the food crops cultivated there – especially fruit. I am addicted to fruit, but at Dokyiwaa I could only shake my head as I beheld the succulent, low-hanging fruits in the compounds and yards of the village. There was a high level of anxiety in the community as people considered the prospects of life in the new location to which they were to be moved from their ancestral land. AngloGold Ashanti made the decision to relocate the community and gave them three months in which to move.[33] They also decided on the value of the compensation to be paid. Some community people swore not to relocate as the compensation was not enough to pay for the move and to survive. The paradox is that staying behind meant certain death.

The area also has a level of malaria higher than the national average. Diseases are attributable to the heavy metals associated with the industry. These include cadmium, lead, copper, mercury, nickel, zinc and manganese. Prevalent health challenges include malaria, respiratory infections, eye infections, reproductive health problems, hypertension and blood disorders.

It would not console the poor community of Dokyiwaa to know that they are not alone in suffering from the impacts of gold mines in Africa. A study of two mines in Tanzania revealed equally worrisome pollution levels. Tests conducted by the Norwegian University of Life Sciences, together with the University of Dar es Salaam, mapped concentrations of trace metal in soil, sediments and waters in the areas close to the Geita Gold Mine and the North Mara Gold Mine in north-west Tanzania.

Tanzania's North Mara Gold Mine (Barrick Gold) is located near Tarime town in the Mara district east of Lake Victoria. For some time, local people suspected the tailing dam was leaking and were afraid that the area was contaminated, while Barrick alleged that people were stealing the lining of the tailing dam and destroying pipes. A major spill occurred here in May 2009.

The academic study found very high levels of arsenic, cadmium, cobalt, copper, chrome, nickel and zinc in the area around the spill. The environment was severely contaminated and, according to the report:

> [A]rsenic (As) contents in the most contaminated water sample were one to two orders of magnitude higher than the WHO drinking water guidelines (10 µg/l). The most extreme water sample contained 8449 µg/l, which is about half a lethal dose of arsenic for a human. Currently, the WHO guideline for maximum As concentration in drinking water is 10 µg/l, but this standard is set to be lowered to 5 µg/l in the near future.[34]

Soil samples showed high levels of arsenic at the spill site and higher than normal levels of other minerals. It is undeniable that the mine is seriously affecting the environment and public health in North Mara.

The area around the AngloGold Ashanti gold mine at Geita was also examined. The scientists noted that although there were higher levels of arsenic and other metals in samples taken, they were found to be within WHO drinking water standards. However, they further noted that there were higher levels in sediments, probably suggesting an unreported pollution incident in the past. People with skin problems consistent with arsenic poisoning were found in both Geita and North Mara.[35]

The global economic crisis and the tendency of mining companies to seek the most liberal tax regimes pose special challenges to governments and people in Africa. In a bid to improve its infrastructure, Zambia's government sought to improve on the tax regime in the mining sector. Some of the measures reflecting foreign pressure included the scrapping of a 25 per cent windfall tax which had been introduced in 2008. In that same year the mining industry kicked and screamed as the government introduced a new variable profit tax, which resulted in raising mining taxes from 31 to 47 per cent. In a move to appease the mining industry, President Rupiah Banda pledged readiness to 're-engage mining companies through dialogue in resolving any of the outstanding (tax) issues, which may have arisen'.[36] Rates were again cut.

Reduced revenue due to the 2008 crash in the copper price, among other factors, made it impossible for Lusaka to deliver on educational, health and social programmes. It also meant lower production in the mines and a resulting cutback on jobs. A total of 10,000 jobs were said to have been lost as a result of the crisis, with a reduction in personal income tax takings from 375bn Zambian kwacha in 2007/08 to 82.2bn kwacha in 2009/10. According to the Zambia Revenue Authority (ZRA) Commissioner-General Chriticles Mwansa, 'mineral royalty revenue fell from over 41bn kwacha ($8.88mn) in August 2008 to 15.2 billion in March 2009'.[37] With this sort of pressure, governments can get desperate. Sixty-three per cent of Zambia's export earnings come through the mines. Overall, while Zambia is currently being promoted as an African success story due to the high levels of resource-directed foreign investment, the everyday reality is quite different for ordinary Zambians. It is a cycle that replicates across much of Africa.

Africa loses a minimum of $148bn each year, four times the amount brought into the continent through foreign aid, and 60 per cent of this is attributed to capital flight due to corporate mispricing of resources extracted from the continent. The obvious solution to this drain of resources is to plug the hole, but the preferred means of tackling the problem has been the provision of more loans and grants accompanied by conditionalities that undermine development.

The Extractive Industries Transparency Initiative (EITI) gives participating corporations such as Shell, Chevron, Vale, BHP Billiton and Anglo American cause to cheer because they can beat their chests and say they paid so much to governments, who cannot say what happened to the funds. The EITI tends to portray the resource-extracting companies as clean while the host governments are dirty and have to struggle to achieve the status of being EITI compliant. By the last quarter of 2010, there were 24 countries[38] in the EITI circuit and more than half of them were African. These include Tanzania, Gabon, Cameroon, DRC, Chad, Mali, Mauritania, Sierra Leone, and Burkina Faso. It is interesting to note that the compliant countries in Africa so far are Niger, Nigeria and Liberia.

The EITI logic, according to Khadija Sharife, is that, 'so long as there is disclosure of cash payments within national boundaries, transparency will act as a natural sanction – diminishing the potential for, and realisation of, corruption. It is a logic that appears to bank on political or "demand-side" corruption, chiefly innate to the developing country's character — with corporations simply "going along" with the system — a kind of "when in Rome" response.' Sharife argues that 'the EITI theory is vastly different from the reality and has more to do with corporate and "first world" country supply-side corruption.'[39]

When Zambia published its EITI report showing payments received from mining companies for 2008 it became one of the most recent countries to make such a publication. It is notable that the image of corruption hangs over countries such as Zambia and scant attention is paid to the corporate tax dodging by mining companies, acts that line the pockets of middlemen in countries with coded bank accounts while depriving Zambia of hundreds of millions of dollars each year.

Zambia's first EITI report – investigating revenues remitted in 2008 – revealed that the government received $463mn in payments from mining companies.[40] The report noted that unresolved discrepancies amounting to $66mn could not be accounted for. That same year, much of Zambia's copper exports, of which almost half were destined for Switzerland, never reached their destination. Paradoxically, the Financial Action Task Force (FATF), referring to Zambia's copper data for the same year, claimed:

> Christian Aid research (p.23) has shown that half the country's copper exports in 2008 were marked at customs for Switzerland. Had Zambia received the export prices that Switzerland did, GDP would have been closer to $25bn than the $14bn that was recorded. That same year, the World Bank records GDP per capita of $1,140 but most recent data (2004) showing nearly two thirds of the population living on less than $1.25 per day.[41]

The extractive sector in Africa has succeeded in making itself seen as the answer to the economic travails of the continent. However, for a country like Zambia, according to Sharife, the extractive corporations:

> generate just 2.2 per cent of revenue collected by Zambian authorities, with the bigger percentage of tax derived from

withheld taxes paid by workers. This is thanks to a particularly nifty boutique tax product called Total Tax Contribution, created by auditing firm PricewaterhouseCoopers, which helps corporations avoid taxation. Zambia's government acknowledged that the country missed cashing in on the 2004–2008 commodity boom, when copper prices more than tripled. But companies like MCM don't have to pay the new royalty rates of three per cent – as 20 year stability clauses from secretive development agreements issued soon after privatisation provided the company with arguably the world's lowest royalty rate at 0.6 per cent. This agreement will remain in force until the year 2020. Worse still, had these agreements not been leaked, it would never have come to light that corporate tax rates were effectively zero, thanks to deferments and royalties.[42]

But this need not always be the case as revealed by the presidential win in September 2011 of long-time opposition leader – and vociferous China critic – Michael Sata. At the Chambishi Copper Mines, mine operator Hedges Mwaba received two different payslips in preparation for the outcome of the election:

'It looks like the Chinese had prepared for any outcome of the election by printing two pay slips for us for the month of September,' said Mwaba to the *Christian Science Monitor*. 'If the incumbent Movement for Multi party Democracy (MMD) [incumbent President Rupiah Banda] had won the presidential election, we would have been paid old meagre salaries. But we got almost double the money because the opposition Patriotic Front led by Michael Sata won the election.[43]

An examination of the economic set-ups that facilitate the avoidance of tax points at the so-called tax havens. Significantly, the UK, a major backer of the EITI, is host to over 50 per cent of the world's tax havens when we count those located in British Crown Dependencies, in British Overseas Territories and in members of the Commonwealth. When members of the Organisation for Economic Development and Cooperation (OECD) are added to the equation we see that the beneficiaries of Africa's impoverishment are actually the most powerful nations.[44]

But tax havens also exist in Africa. African countries nurturing and benefiting from being tax havens include Ghana, Botswana,

Seychelles, Liberia, Djibouti and Mauritius. These serve as conduits for tax avoidance; their economic systems are structured around unearned resource revenues.[45]

Mauritius provides a hub for 'round-tripping': Indians park their cash there and later reinvest it in India on a tax-free basis. Through this arrangement, Mauritius receives $39bn-worth[46] of investment from India, amounting to almost 50 per cent of its total investment flows. The country hosts several dozens of multinational subsidiaries – some of which are accounting firms – offering tax-'planning' products. The OECD only stopped classifying Mauritius among countries listed as 'uncooperative tax havens' after the ratification of 'bilateral tax arrangements' related to 'suspected tax evasion' as well as 'data on request only'.

The Tax Justice Network awarded Mauritius an 'opacity' score of 96,[47] a mere four points away from total opacity. This was due to banking secrecy and insufficient compliance with international regulatory requirements, as well as not maintaining company ownership in official documents.

In the United States, transnational corporations listed on the US Securities and Exchange Commission (SEC) spent a lot of energy in 2011 inserting clauses in a new financial reform proposal to ensure that they could continue operations on the well-worn path of double 'transparency' standards. The reform requires that mining, oil and gas companies which trade their shares on the American stock exchange must issue an annual report detailing the 'type and total amount' of payments they make to foreign governments. Writing to the SEC, the American Petroleum Institute (28 January 2011) and the mining giant Rio Tinto (2 March 2011) laid out reasons why there should be escape hatch clauses which would allow them to choose which laws to obey and which to break – the laws of either the United States or the countries in which they operate.

There are countries in Africa where domestic laws classify revenues obtained from the extractive sector as 'state secrets'. On account of this, Rio Tinto argued:

> We believe that there should be an exemption, if such reporting would violate, or may reasonably be deemed to violate, host country laws. The proposed disclosures, for instance, may well

constitute state secrets for a project in a specific country. The issuer should not be forced to choose between which law it will violate – the U.S. or the host country laws.[48]

The American Petroleum Institute focused at length on what they saw as the 'potential for competitive harm' in the reform. Their argument is that disclosures would enable competitors who are not listed on the SEC to use such disclosed figures to undercut or outmanoeuvre them in bids. They also opine that disclosures may endanger the lives of workers in the sector, as 'Energy companies have already experienced numerous incidents where facilities have been sabotaged, operations disrupted or employees endangered by those who oppose the host country government or energy development.'

Interestingly, corporations do not have qualms about the EITI provision that requires the private sector to disclose payments made to governments and governments to disclose payments received. According to Rio Tinto, 'Because the EITI also encompasses disclosure by governments of payments they receive from companies, we believe it is more effective than the proposed rules at improving governance and eliminating corruption in both the private and public sectors. Therefore, we urge the Commission to follow the EITI principles to the fullest extent possible.'

The American Petroleum Institute recommends 'that the Commission require issuers to report payments based upon amount actually paid by the issuer to the government entity (as opposed to the issuer's net share of the payment), consistent with the EITI practices'. Furthermore, the API does not 'believe it is necessary for the rules to specifically list other types of fees that would be subject to disclosure. We note that fees related to entry into, or retention of, licenses or concessions can be competitively sensitive information.'

Reform would therefore be not just a threat to the companies in the sector, who are used to having smooth rides over corrupt waters in certain countries. It should also worry governments in Africa and elsewhere that are not open to disclosures of payments made in the sector. Fingers have been pointed at many resource-rich countries. For example, American Petroleum Institute members 'can confirm to the Commission that disclosure of revenue payments made to foreign governments or companies owned by foreign governments are prohibited for the following countries: Cameroon, China, Qatar, and Angola.'

The extractive sector companies require the goodwill of host governments to operate in their countries mainly because of the huge environmental and human rights abuses that accompany their actions. Providing escape routes and bending the rules to suit their practices helps entrench double standards and aids destructive extraction. It is therefore not surprising that the government of Zambia was unhappy about opposition politicians' efforts to block the handing over of the closed Luanshya Copper Mines (LCM) to China's NFC Africa.

The mine had closed in December 2008 after making serious losses. Was the government's stance toward China an attempt to restrain the latter's unchecked, growing influence? Chinese corporations have poor safety records and also pay low wages. The Chinese safety record in Zambia includes a 2005 mine accident at the Chambishi mine in which 49 miners died. A Chinese company, NCFA, took over the mine in 2003. The Chinese influence is also a source of resentment in Zambia due to their involvement in clothing and other light manufacturing sectors. These sectors have deindustrialised across Africa due to East Asian imports.[49]

Talking about Chinese operations in Africa can be sensitive business. After years of exploitation by Western governments and institutions, some people welcome Chinese competition, in part because the funds they bring in are not accompanied by political and environmental conditionalities. But those conditionalities are, in some cases, reforms that we have sought in order to protect our societies and ecologies from the resource curse. China is, therefore, sometimes a more serious threat than even the worst Western multinational corporation.

For example, in October 2009, before the blood of over 150 civilians mowed down in cold blood by the military junta in Guinea could dry, Chinese investors became the junta's strategic partners for mining projects. The Chinese invested over $7bn in infrastructure in the poverty-wracked nation at a time when the world was aghast at the massacre of unarmed pro-democracy activists who had gathered in a stadium. It is estimated that the country may have the largest deposits of aluminium ore and bauxite in the world.[50]

The Horn of Africa is an area that is especially rich in natural resources as well as tyrannical and failed states. Indeed, Somalia has been so ravaged that it occupies a special place in the understanding of what makes a nation in today's world. Close to Somalia are Ethiopia and Eritrea. There are many reasons for the conflicts and tension between Ethiopia and Eritrea, both of which are repressive dictatorships. Although there is no open war between the two, both countries amass troops along their common border. The last major conflict was in 1998–2000 when tens of thousands of military personnel died in a battle for a dusty strip of land. It is estimated that 10 per cent of the 4.5 million citizens of Eritrea serve in their country's armed forces, making it one of the most militarised countries in the world. One of the least developed countries in the world, it nevertheless spends over 6 per cent of GDP on its war machinery.

On the western border of Eritrea sits the town of Bisha where Nevsun, a Canadian mining company, found a stack of high-grade gold atop copper, silver and zinc deposits at depths of up to 450m. Nevsun is investing $250mn to build a mine from where they will exploit this spectacular find. The company operating here is the Bisha Mining Share Company, a joint venture between Nevsun and the Eritrean National Mining Corporation. Nevsun owns a 60 per cent share of the company. The company dreams of extracting a hefty 28 tonnes of gold, nearly 280 tonnes of silver, almost 340,000 tonnes of copper and more than 450,000 tonnes of zinc over its projected 10-year life.[51]

In another tragic example, Tanzania is infamous for the severe human rights abuses in the mining sector perpetuated against its citizens in order to pave the way for corporate takeover. The infamy arose from the conflict between local livelihoods and the quest for profit by big industry. The odd fact here is that the government displaced small-scale miners who were known to bring in more government revenue than the big miners would. Gold was discovered in the Bulyanhulu area of Tanzania in the mid 1970s. Mining was carried out primarily by an estimated 30,000–40,000 small-scale miners, who by the early 1990s were bringing an estimated $30mn annually to the national coffers. The highest incomes in the region were in the mining area. The miners paid taxes and invested in their local economies.

All these were jolted when in the mid-1990s, Sutton Resources Inc. of Canada was given the rights to the area and obtained a permanent injunction, blocking small-scale miners, who were now considered to be illegal squatters. Using paramilitary troops, small-scale miners were evacuated from the area without any compensation. As many as 52 miners were buried alive as Sutton's graders filled up mining pits and tunnels. Triumphant on the caked blood of artisanal miners, Sutton was purchased by the giant Barrick Gold Corporation of Toronto, and began to exploit the continent's largest gold deposits. Early estimates were that the corporation would earn $60–75mn in profits per year while paying taxes and royalties worth $5mn to the Tanzanian government and carrying out social service projects worth another $10mn. Without counting the socio-economic and environmental impacts and loss of local livelihoods, we do not need an economist to unveil the scam powered here by the machinery of deregulation.[52]

Enthusiasts claim that mining companies in Tanzania will bring in taxes amounting to about $300mn (about Sh390bn) annually by 2014 compared to the current level of about $100mn. But critics accuse the mining companies of tax avoidance and evasion, which has cost the nation about $400mn in the past decade.[53]

The burial of small-scale miners also occurred in Ghana in 2009. But for the quick intervention by other community folk, these miners would have been entombed alive. The National Coalition on Mining (NCOM) from Chirano, Obuasi, Kenyasi, New Abirim, Bibiani, New Atuabo near Tarkwa, Prestea, Mpatuom, and Accra condemned AngloGold Ashanti for the attempted murder of small-scale miners (*galamseyers*):

> On Tuesday August 11th, 2009, we learnt with shock and deep regret that the security personnel of AngloGold-Ashanti allegedly decided to bury alive 40 small-scale miners at one of the company's abandoned pits at Tom Collins also known as (Blacks Pit) near Obuasi the Municipal capital in Ashanti Region of Ghana. We view the action of AngloGold-Ashanti as barbaric, high-handedness, torture and attempted murder of citizens.[54]

Transnational corporations profit both through these kinds of immoral production relations, as well as through straight corruption and capital flight, for there are many African leaders who stash away funds – often straight bribes – in coded bank accounts overseas and whose personal wealth is said to be equal to their national debt. Some, who may actually forget their passwords and die off, leave the funds to float away straight into bank profits.

With more talk about transparency and accountability, has the situation changed? Not everyone thinks so. 'The rate of corruption does not appear to fall even when the level of national revenues fall ... because corrupt politicians ensure they keep their lives of conspicuous consumption.'[55]

I will conclude this chapter by looking at what oil means to the current and political future of Uganda, where tensions have risen in recent times and where walking to work became a simple act of dissent.

Dream find in Uganda

In 2006, multinationals made a long-suspected oil discovery in Uganda. The black gold reserves in the Albertine region comprise some 23,000sq km along the border of the DRC, potentially catapulting the region into one of the world's 50 top oil producers. Two billion barrels were discovered with only 25 per cent of the area explored. By early 2012, production is targeted at 220,000 barrels per day. In 2009, when 800 million barrels were confirmed, Tullow Oil, one of the multinationals involved, pegged annual revenue at $2bn. In that same year, *The Guardian* newspaper in the UK reported the hopeful statement of Margeret Ayuro, a mother of eight from Abule village in Uganda: 'I believe God will make the government help us, since he has opened our eyes to be able to see that oil.'[56]

Like Ayuro, Museveni credited the discovery of oil to the glory of God, holding a national prayer ceremony where he publicly thanked the creator for developing 'the rift valley some 25 million years ago'.[57] Museveni may as well have been thanking himself. In September 2005, just prior to the impending discovery, as oil companies like scrambling for greater and greater concessions, Museveni – holding power for 25 years – was blessed by Uganda's parliament, which altered the constitution so as to remove term

43

limits for the president. This enabled Museveni to run multiparty elections, the first since 1980, which he won with 59 per cent of the vote. But opposition leader Kizza Besigye, with 37 per cent of the vote, challenged the results in Uganda's Supreme Court, which acknowledged serious irregularities, though these were allegedly not sufficiently serious to overturn the results.

Earlier, Besigye was arrested by Museveni's government for treason, among other crimes, and then released by the High Court when the Netherlands and other foreign donors withdrew funding. Critics claim that Museveni's turn toward multiparty elections was prompted by the actions of donor countries, indicating the power that foreign funders wield over African dictators. The US considers Museveni's government as a key regional ally. In early 2011, in anticipation of future oil revenue, Museveni purchased at least $744mn in Russian arms with at least eight fighter jets.[58] Museveni's son, Colonel Muhoozi Kainerugaba, a Fort Leavenworth and Sandhurst-trained leader, is the head of the country's Special Forces, guarding the oil. In 2011 he also led violent crackdowns on civilians peacefully protesting the high prices of food.

'People are desperate,' said Ugandan attorney and Harvard Law PhD Emmanuel Bagenda:

> Even sugar – basics – are beyond the reach of normal people. The amendment of the Constitution is key to the problem: as long as the current patronage system remains, those who are in power cannot be held accountable. We talk about justice, but where is it being implemented? Why do donors continue funding this system?[59]

Part 2
The scramble and the grabbing

The wheels of progress

Climbing the trails of tales
We glean doubtless transitions
Folks rising on survival trails
Seas rising on suicide runs
Monster monuments to bottomless greed
Ruins, debts and disappearing wealth
Fossils
In the matrix of invisible smokescreens
We walk blind!
And dance to broken drums[1]

THE WHEELS OF progress often roll forward, but unevenly with bursts of speed and then periods of stagnation, and as we see in Africa, sometimes they roll backwards. But in considering that the entire world has rolled to the edge of a sustainability cliff with economic and environmental crises soon to tip us over, survival may require that the wheel of forward 'progress' begins to roll back a little bit. Otherwise the only safe berth may be another planet.

Humans may have taken their first steps in Africa, but their sudden flight through technological development took place in the North. That flight was made possible by the development of ideas and strategies that ensured the comfort of the elite and their business concerns. Major waterworks and other infrastructure, which fed the need for further power and territory, facilitated the ancient centralised societies of China and Rome. Over time, the proverbial commoners in the North and the majority of peoples in the South got sucked into unequal relationships. As time went by, the emerging systems of exploitation of the silent have come to be accepted as the norm, and those who raise voices against them are sometimes termed extremists.

Without having to reach too far back, we can step into the era of pre-colonial adventures. Men risked their limbs and lives as they sailed into uncharted territories, seeking out unknown treasures and opening up markets in which they could exert the power of monopoly. Those who returned to the imperial courts with tokens of their exploits were routinely rewarded with titles and an assortment of decorations. Thus the South provided the grounds for vicious games of exploration and appropriation. The scramble for Africa in the 19th century saw European powers falling over each other as they carved up the continent. There were resistances, but there were more often massacres. We read a bit from Thomas Pakenham:

> Soon the Maxim gun – not the cross – became the symbol of the age in Africa (though in practice the wretched thing jammed, and the magazine rifle did the job better). Most of the battles were cruelly one-sided (but not for the British against the Boers, or for the Italians against the Abyssinians). At Omdurman, British officers counted 10,000 Sudanese dead or dying in the sand. They made no effort to help the 15,000 wounded.[2]

The incursions into Latin America and Asia and the destruction of civilisations there in the service of imperial power are well documented. After the seasons of conquest and pillage, including the slave trade and colonialism, where was the North headed?

We see a two-pronged movement. One track helps to secure the consumerist bent of its society and the second prong has devised means of keeping the South under its thumb for the purpose of extracting the raw materials needed to feed the deep throats of consumerism. Could this happen without collaborators in the South? Could it be that those collaborators never knew they were working against their home region's best interests? The answer here would be yes and no. There are those genuinely convinced that any business is good business as long as it pads their wallets. And there are others who do not have a clue that they are part of the machinery keeping their peoples in penury. This second group is made up of people assimilated by the system and others. In the entire matrix of the exploiters and the exploited we see that the North can be found in the South and the South can be found in the North. But we are concerned here with the broader and more popular usage of the terms.

We have seen in recent times how the public gets manipulated into backing vicious state actions. The starkest example is the manner in which the United States went into the war against Iraq, using as cover the despicable 9/11 attacks at the World Trade Centre and the Pentagon. The main excuse given, that the Iraqis had stocks of weapons of mass destruction, was a barefaced lie right from the beginning and the officials pushing the lie knew that it was so. Yet they led the American public to believe that if nothing was done, and quickly too, there would be more dastardly attacks at home.

To further heighten the sense of insecurity among its people, the government constructed a visual image of an axis of evil and hung the spectre as a totem on every heart – and went right ahead to sacrifice many lives in needless battles in the streets of Baghdad, Basra and others. After hundreds of thousands of Iraqi and thousands of American and allied lives were lost, the world is not any safer than it was before the attacks. Indeed, it can be said that the presidency of George W. Bush made the world more unsafe by the day. That sense of insecurity was a great opportunity for what former US President Dwight Eisenhower called the military-industrial complex to roll out more weapons and for military contractors to have their greatest boom season ever.

One of the well-kept secrets of the war in Iraq is that the smart bombs used to devastate that nation are not as smart as they are advertised to be. A commentator in the video documentary *Why We Fight* alleges: 'During the first 6 months of the Iraq war, 50 precision airstrikes were conducted against the Iraqi leadership. Of these strikes, none hit its intended target.'[3] Civilians were killed in their thousands.

The wars are hinged on the storyline that the troops are fighting to defend a way of life that must not be altered. This includes spreading democracy and liberty. Past presidents expressed the same doctrine in different words. Whoever does not accept this philosophy is seen as an enemy of peace and must be hunted down, especially if they sit, as Pentagon war architect and later World Bank president Paul Wolfowitz put it in relation to Iraq, 'atop a sea of oil'.

Perhaps it is idealism for preserving freedom that makes soldiers shoot at people they know nothing about. The military provides jobs to some who find it tough securing employment in depressed local conditions. And with the newly invented war support systems, soldiers only have to eat and then fight. Everything else that is not a direct combat function is farmed out to private companies such as Halliburton. In this scenario, globalisation makes it easy to find workers from the migrant pools who are willing to handle the tasks for the troops in any sort of condition. And so, as Pratap Chaterjee reports, the soldiers happily fight while civilians handle their laundry, sweep their rooms, cook their foods and generally:

> provide a dazzling array of creature comforts that convinced reluctant U.S. soldiers to continue to sign up to fight in Iraq – from all-you-can-eat dinner buffets to Burger King and Pizza Hut on demand, as well as hot showers and an endless supply of video games – mimicking their lifestyle back home, except that in Iraq, the soldiers didn't have to clean up after themselves.[4]

The way to a man's heart may be the stomach, after all.

The North has relentlessly pressed for the life of ease. Everything is aimed to make life easier and if it were possible travel could become a push button affair – where all you have to do is press a button, dematerialise and reappear at your chosen destination. Or better still you could press a button and your destination would arrive at your doorstep. We will perhaps get there.

For some time the North could be said to have had everything under its control. That happened with regard to crude oil as a major energy source and driver of economic development. The United States, for instance, was self sufficient in oil. The stuff was cheap, and there was little thought given to the possibility of change in the equation. In fact, in the early 1960s crude oil cost less than \$2 per barrel and the US was able to meet up to 70 per cent of its needs from domestic production. However, things began to change from 1970 when US oil production began to dip.

The Organisation of Petroleum Exporting Countries (OPEC) had come into existence and in 1973 deliberately began engineering oil supplies and pricing, driving up oil prices by about 70 per cent, hitting the unprecedented level of more than $5 per barrel.

By 1981 oil prices were close to $40 per barrel, setting off alarm bells in the North. This upsurge in oil prices pushed the North to embark on serious investment in alternative energy sources, including nuclear power, as well as the development of more oilfields in non-OPEC countries. It also brought to the fore the need for more efficiency in energy utilisation. Oil prices fell to about $10 in 1986.

When Saddam Hussein invaded Kuwait in August 1990 he may have had his sights on the annexation of the Kuwaiti oilfields. The invasion jolted the oil market, and the price went up to more than $30 per barrel. A perhaps bigger jolt was to the big consumer nations such as the US, who feared what might be next on the Iraqi president's list. Would he attack Saudi Arabia? What would that mean to the world of crude oil supply? Saddam had to be stopped. He was – when he swung on the gallows.

The price of a barrel of oil tumbled to $10 in 1998 due to lack of agreement among OPEC members over production levels. Over the years there have been efforts to break up the OPEC ranks or to diminish its membership through splits in the organisation, but no such major quakes have happened so far.

Although the price of oil has risen beyond what could have been speculated a few decades ago, the commodity still retains a commanding place in the energy mix. Besides, many everyday products that we depend on today have petroleum inputs in one form or the other – not counting the energy with which they were manufactured. Look around you and see how many items contain petroleum products: the cars we drive, plastic bottles, building materials, foods and their packaging, electronic equipment, textile materials and even medicine. Petroleum is so ubiquitous that we take it for granted, or better put, we don't even see it. It is what drives the machines of war. It also drives other policies of state.

Addiction to oil is deep. Going 'cold turkey' would manifest severe withdrawal syndromes for consumers as well as suppliers. Consumers are hooked on the stuff and the suppliers are hooked on petrodollars. Both sides prefer that oil supplies run forever.

But unfortunately, oil is a finite resource and must run out or become less reachable, irrespective of technological innovations. And that should be a good thing, at least for the climate, and also for the oilfields of the South, which can also be termed killing fields. With oil drying up, it might make sense to earn money from pollution if you cannot stop polluting. This appears to be the driving force behind carbon trading, one of the many false solutions to climate change. This solution follows a familiar trajectory: placing reliance on the virtues of the free market and, conversely, rejecting anything that has elements of state or popular control.

With faith in the market as the cure all, the North relentlessly pushes for commodification and, as some say, financialisation of everything, including life. This has already manifested itself in the patenting of life forms in efforts to appropriate the creation of seeds and animal life that never had their origins in the tinkerer's test tube. This is the turf of those engaged in genetic engineering. People have gone ahead to commodify carbon and throw up an army of carbon market speculators.

Today swaths of forests are traded because they are suddenly found to hold tons of carbon and investors must be encouraged to pay and by so doing own these trees for the sake of the climate. So the North is leading the way in enthroning briefcase approaches to tackling the climate crisis. This carbon investment is meant to sink the chainsaws of deforestation. It does not matter how much of the world's resources are chewed up and how polluting fossil-fuel-driven industry is.

With the market doctrine firmly in policymakers' thoughts, everything becomes a candidate for the auction block. Water is already privatised to a large extent and probably for good reason. In the past people depended on natural water bodies for potable water. Today rivers have been bought up for private use and others have been plied with toxic effluents from industries as well as urban sewage. In the United States for instance, a couple of decades ago, everyone depended on public water fountains if they needed to slake their thirst in public. These were so useful that the fountains became points for racial segregation. Writing about the rise of bottled water, US commentator Saul Landau states:

In my youth, I don't recall people drinking water from plastic bottles. We used public fountains. Before privatisation, bottled water couldn't have competed with tap water. I attribute the triumph of expensive bottled (contents unknown) over cheap tap water (contents monitored regularly) to the steady decline of the political alliance between the poor majority and the government.[5]

All these have happened because people have accepted the doctrine that the market holds the key to progress, government is the problem and the private sector has all the brains and skills in the world to make things right. When the world faced the economic meltdown of 2008–09, it was the same derided public sector that provided safety nets for the private sector whose greed and lack of transparency in their dealings had precipitated the crisis in the first place. Arthur Schlesinger Jr said that getting government off the back of business simply means putting business on the back of government.[6]

The enthronement of the market as king has facilitated the removal of trade barriers perceived by governments as protectionist while reinforcing rules that would make it impossible for the disadvantaged nations to gain access to the markets of the powerful. The particular consequence of this can be seen in Africa where some countries depend largely on agricultural exports for their foreign exchange earnings. In the late 1990s the European Union banned fish imports from Kenya, Uganda, Mozambique and Tanzania because of concerns over sanitary standards and control systems. Through that measure, Ugandan fishermen lost $37mn while those in Tanzania suffered an 80 per cent erosion of their income.

According to the Word Bank, African exporters of cereals, fruits, vegetables and nuts face estimated costs of about $670mn annually to meet the EU's requirements over aflatoxins, because these are stricter than those of the joint FAO/WHO Expert Committee on Food Additives.[7] Meanwhile more African farmers turned to flower cultivation for export to Europe and other markets from the 1990s. Horticulture experienced a leap in a country such as Kenya, where the fertile Lake Naivasha area attracted many such transnational agribusiness/farmers, placing serious pressure on the water available for the production of food crops and threatening the food security of the region. The flowers produced here are for the pleasure of wealthy consumers thousands of miles away.

That this market is drawing farmers away from cultivating much-needed food crops is not seen as a problem, because when some move into the flower business, others will occupy the space they left. The reality is that more and more farmers in Africa go into the flower business, rather than feed themselves and others. They cater for the aesthetic needs of decorators, but when times are bad, are hard pressed to fill their own stomachs. Land is seen as a commodity rather than as a place for the production of food and sustenance.

The global food crisis has triggered land grabs that could not have been imagined before. Nations with cash but not much arable land are buying up chunks of land in Africa, mainly for the purpose of cultivating crops to be used in biofuels production or to grow food crops that get repatriated to the country that bought the land. The logic is that of the market: you have land; I have cash. We are both happy. But there is far less land left for cultivation of food crops for local consumption and the diversion of labour, water and other inputs from producing food for local communities will have an adverse impact on food production.

The debate on overseas development aid has been going on for a long time, but very interesting issues have come out of late. Some of these have been whispered in the campaign meetings of civil society groups but are now being discussed in the open. The underlying point is that development aid or any other aid that flows into Africa can hardly be acts of benevolence. The North sometimes uses aid as a wedge to force particular policy options on aid recipients. Dambisa Moyo,[8] Robert Calderisi[9] and others have represented the ideologies of the World Bank and the International Monetary Fund (IMF), respectively.

One problem with Moyo's argument is the premise that the West made the mistake of 'giving something for nothing'.[10] Yet writing about the oil crisis of the early 1970s, Moyo admits:

> As oil prices soared, oil-exporting countries deposited the additional cash with international banks, which in turn eagerly sought to lend this money to the developing world ... The wall of freely supplied money led to extremely low, and even

negative, real interest rates, and encouraged many poorer economies to start borrowing even more in order to repay previous debts.[11]

We see from the above that some poor but rich countries such as Nigeria actually got into the debt trap by borrowing their own money. Moyo correctly observes that with the mass of staff people involved in running the aid machinery, there is an in-built pressure to keep it running. And even where donors would like to stop lending or giving, and would like not to give to corrupt regimes, they are unable to agree on who is corrupt and who is not. This does not occur just because of clientelistic relations between the Western donors and favoured dictators. Another basic reason why donors find it hard to halt this form of aid-as-imperialism is found in the reality that aid giving is a lucrative business for the donors. According to Patrick Bond:

> Because of vast wastage associated with the aid bureaucracy, tied aid, as well as other 'phantom' aspects such as debt relief, a further correction to the statistics can be made. Globally, according to Action Aid, total official aid of $69 billion in 2003 was reduced to 'real' aid to poor people of just $27 billion. About one seventh (14 per cent) of the purported aid – better considered 'phantom aid' – includes 'debt relief' which rose from around $1.5 billion in 2000 to more than $6 billion in 2003… [T]he debt relief was provided in such a way as to deepen not lessen dependence and Northern control of Africa. Other phantom aid components include the transaction and administrative costs of paying out aid funds (14 per cent). Technical assistance by Northern experts accounted for a fifth of aid; as noted, water privatisation advice by Britain's Adam Smith Institute is an example of how such donor assistance does yet more damage to the African state and society.[12]

So aid equally benefits the donors in terms of the amount designated for consultancy and equipment that must be procured from the donor nation. In 2004, a whopping 40 per cent of aid given to Africa went to pay international consultants connected to the aid.[13] From this premise it could be argued that aid is nothing much more than a hypocritical, self-serving form of Western corruption. It permits the donor to stand back and exclaim, 'We have done our bit, but they the African people just will not change!' And then it opens the doors for the drawing up of new strategies for doing the same things to the same ends. In his book, Calderisi proposes ten ways for changing Africa via aid:

- Introduce mechanisms for tracing and recovering public funds
- Require all heads of state, ministers, and senior officials to open their bank accounts to public scrutiny
- Cut direct aid to individual countries in half – essentially reducing rather than increasing aid
- Focus direct aid on four to five countries that are serious about reducing poverty
- Require all countries to hold internationally supervised elections
- Promote other aspects of democracy, including a free press and an independent judiciary
- Supervise the running of Africa's schools and HIV/AIDS programs
- Establish citizen review groups to oversee government policy and aid agreements.
- Put more emphasis on infrastructure and regional links
- Merge the World Bank, IMF and the United Nations Development Programme.[14]

Calderisi argues that the implementation of any one of the ten ways would enhance the lot of African countries and that taken together the continent would literally fly on the path of progress. Of course, most of these could be recommended to any nation on earth, as corruption is not inherently African. Even the head of the IMF, Christine Lagard, was investigated for crony capitalism associated with French politics when she was finance minister.

But some of Calderisi's ideas are rather paternalistic, if not neo-colonial. Without saying so outright, he also raises the problem of who would determine that African countries keep on track. Would this be the morphed entity, the merged the World Bank, IMF and the United Nations Development Programmme? Such a merger proposal is a serious indictment of these organisations. They need to be individually re-examined, emptied of orthodox economists, perhaps closed down, and at a minimum realigned to the needs of a truly free world where nations relate in assured dignity and respect.

The HIV/AIDS problem is something that hit Africa literally below the belt, exposing us to the politics of disease and medicine. Apart from the fact that some leaders on the continent, such as South Africa's Thabo Mbeki, did not readily accept that the scourge was an acute problem, others found solace in the role of activists fighting to access life-saving AIDS drugs. According to Bond, the highest-profile aid interventions in recent years were probably in the field of HIV/AIDS treatment. These interventions initially included threats of aid cuts against governments like Nelson Mandela's South Africa, which made provisions for generic medicines production. US president Bill Clinton only backed away from this threat in late 1999 because of sustained popular protest.[15] Bond also notes that George W. Bush in early 2003 promised a \$15bn AIDS programme, then significantly reduced the funding and also refused to provide adequate resources for the UN Global Fund to Fight AIDS, TB and malaria, and for a while even prohibited US government financing of generic medicines.[16]

The bottom line of Northern interest in Africa is to keep markets open for the extraction of raw materials and for the sale of finished products. Advice from international advisers routinely aims at forcing countries to comply. The pliers used to pry open markets are sometimes hidden in gloves with the labels of support for good governance and social investment. Who would argue against these? Countries compete to receive these labels under, for example, the Millennium Challenge Account (MCA) introduced in the Bush era. Attainment of these seemingly benign conditionalities are not set by the competing countries, of course. With 39 African countries competing in 2004 in a field of 74 eligible low-income countries, only eight from Africa made it: Benin, Cape Verde, Ghana, Lesotho, Madagascar, Mali, Mozambique and Senegal. Among the agencies that examined these poor countries is the World Bank.

As set out in *Looting Africa*,[17] the criteria for funding these countries' aid programmes fall into three categories:

Ruling justly: Based on Freedom House rankings of civil liberties and political rights as well as World Bank Institute indices on accountability, governance and control of corruption.
Economic freedom: Determined by credit ratings, inflation rates, business start-up times, trade policies and regulatory regimes as measured by such institutions as the World Bank, the International Monetary Fund and the Heritage Foundation Index of Economic Freedom.
Investment in people: Gauged according to public expenditure on health and primary education, immunisation rates and primary school completion rates as recorded by the national governments, the World Health Organisation and the UN.[18]

Besides aid that promotes neoliberal doctrines and policies, war is also used as an instrument for the implantation of 'democracy'. It is generally believed that if you repeat a thing often enough you may begin to believe it – irrespective of whether it is right or false. Before and after the war on Iraq, the claim was that Saddam Hussein had a stock of weapons of mass destruction (WMD) and that the elimination of that stock was a critical part of the war against terror. Right? When it was clear that the chase after WMD was phoney, some top US politicians persisted in selling that notion. For instance, in January 2005, Bill Frist, Senate majority leader, said of Iraq: 'dangerous weapons proliferation must be stopped. Terrorist organisations must be destroyed.'[19]

Military might has also been viciously used to shatter nations, as seen in Somalia. The many wars in Africa, especially in mineral-rich areas such as the Great Lakes of Central Africa and the rich fields of Sierra Leone and Liberia, can all be traced to prodding that came from extractive corporations and from home governments that would not call them to account. Evidence of a revolving door between state and corporate power was provided in September 2011 when it was revealed that the British government's Mark Allen supported torture and rendition in Libya on behalf of Washington, coinciding with his assistance to British Petroleum, which then made him a senior adviser.

The mix of military and commercial interests reminds us that Northern over-consumption means that to maintain the trajectory of human life based on our current standard of living requires more planets for resource extraction. Perhaps this is why space exploration is intensifying, including the bombing of the moon in October 2009 by the US. Bombing of the moon? Indeed. The US reportedly dropped two rockets on the moon in order to raise a cloud of dust from which scientists were expected to measure and ascertain whether there was moisture on the moon. Places and things you may have thought were in the global commons are gradually being appropriated. Recall that in search of oil Russians planted their flag in the seabed beneath the North Pole. It is obvious that tomorrow's resource wars have already started.

It would appear that the direction is in the North going from fast food to slow food, from moon-banishing city lights to the enjoyment of moonlight. In what direction must the wheel of progress roll?

While we ponder, the nagging question remains why African governments accept strategies and conditionalities they know are harmful to their people and are contrary to the steps they need to take in order to fulfil their roles as governments? Is it because multilateral agencies and international financial institutions must first endorse almost every policy step these governments take? We will examine that further in the next chapter.

The steps of the advisers

Today, awed fingers
Tickle your sides
Ripples of ancient mysteries flash across
Your mystic face
Life tales in your belly
How deep is your heart
That men sought to drain your guts
And drain your blood to
Scrape golden dregs from your belly
Through a straw[1]

AS ADVISERS FROM multilateral agencies and foreign governments step into Africa the shout on their lips is nothing other than, 'Liberalise! Liberalise! Liberalise!' But the response from the lips of the African peoples is, 'Liberate! Liberate! Liberate!' The two slogans do not match.

Trade liberalisation is a core tenet of the structural adjustment programmes (SAPs) of the World Bank and the International Monetary Fund (IMF), and we will look a bit more at this as we consider policies that Africans have had to confront or adopt. Whereas trade liberalisation in theory should open up business for all sides, this has not been the case. Is there something inherently wrong with the idea? Perhaps not. What must be wrong can be traced to the mode of implementation and the hidden hands manipulating so-called free-market forces.

Trade liberalisation has played out very badly in African countries where the policy has forced them to throw their doors wide open. First of all, governments placed emphasis on export crops. Food prices collapsed in Ghana and farmers could not recover their costs. In Kenya, food imports also increased and the resultant food dumping played havoc with local farmers' livelihoods. In Benin, the contest was between the government's focus on cotton cultivation and export, and the farmers' cry for help in food production.

Although these policies were wrong-headed, the response of the financial institutions has been to tinker with labels and names rather than address the substance. This is not incidental. Free-market evangelists are well entrenched in the power structures of the North and keep on hammering the same doctrine. After all, they are not the ones who have to swallow the bitter pills.

Many have placed hope for change in Barack Obama's administration. But with the top economic adviser none other than Lawrence Summers in 2009–10, who really expected any substantial shift in the status quo? As chief economist at the World Bank, he was a strong backer of the SAPs. Mr Summers is also known for hanging on dogmatically to absolute positions, according to Naomi Klein:

Back in 1991, Summers argued that the subject of economics was no longer up for debate: The answers had all been found by men like him. 'The laws of economics are like the laws of engineering,' he said. 'One set of laws works everywhere.' Summers subsequently laid out those laws as the three '-ations': privatisation, stabilisation and liberalisation.[2]

Summers had, the same year, argued that hazardous wastes should be disposed of in Africa:

Just between you and me, shouldn't the World Bank be encouraging more migration of the dirty industries to the LDCs [Lesser Developed Countries]? I can think of three reasons: (1) The measurement of the costs of health-impairing pollution depends on the forgone earnings from increased morbidity and mortality. From this point of view a given amount of health-impairing pollution should be done in the country with the lowest cost, which will be the country with the lowest wages. I think the economic logic behind dumping a load of toxic waste in the lowest-wage country is impeccable and we should face up to that. (2) The costs of pollution are likely to be non-linear as the initial increments of pollution probably have very low cost. I've always thought that under-populated countries in Africa are vastly under-polluted; their air quality is probably vastly inefficiently low [sic] compared to Los Angeles or Mexico City. Only the lamentable facts that so much pollution is generated by non-tradable industries (transport, electrical generation) and that the unit transport costs of solid waste are so high prevent world-welfare-enhancing trade in air pollution and waste. (3) The demand for a clean environment for aesthetic and health reasons is likely to have very high income-elasticity. The concern over an agent that causes a one-in-a-million change in the odds of prostate cancer is obviously going to be much higher in a country where people survive to get prostate cancer than in a country where under-5 mortality is 200 per thousand. Also, much of the concern over industrial atmospheric discharge is about visibility-impairing particulates. These discharges may have very little direct health impact. Clearly trade in goods that embody aesthetic pollution concerns could be welfare-enhancing. While production is mobile the consumption of pretty air is a non-tradable. The problem with the arguments against all of these proposals for more pollution in LDCs (intrinsic rights to certain goods, moral reasons, social concerns, lack of adequate markets, etc) could be turned around and used more or less effectively against every Bank proposal for liberalisation.[3]

With this sort of thinking from a chief economist at the World Bank it can be seen very clearly that pollution does not happen by chance. Poverty and ill health are, apparently, engineered through an ideological mindset that sees as appropriate anything that secures the wealth of the mighty even if it kills or diminishes the capacity of the weak to survive. With this sort of mindset, we see that policies are purposely skewed against Africa. Consider the example of agricultural subsidies: when he was US treasury secretary, Summers continued another set of policies – massive state subsidies for agribusiness. Those subsidies, across Europe, Japan and North America, expose the unfair standards foisted on Africa by the North. It is estimated that such subsidies to farmers amount to almost a billion dollars a day – more than the GDP of sub-Saharan Africa.[4] Subsidies also reveal inequities within the North itself, because it is big agri-business that enjoys the huge subsidies rather than the smallholder farmers in Europe or North America.

Often, there is a tendency to think there are no small-scale farmers in the North. The truth is that they exist and that they provide essential and needed wholesome foods that meet the needs of local communities without having to depend on shipments of foods and products across vast distances. Most of these small-scale farmers utilise labour-intensive, agro-ecological methods and do the environment and local communities and neighbourhoods a whole lot of good.

It may surprise many to learn that the smallest 60 per cent of European farms receive only 10 per cent of the subsidies available in Europe, while the top 2 per cent receive nearly 25 per cent of the subsidies. The ratios are even starker in the US, where 60 per cent of the farmers receive no subsidies at all. Contrast this with the fact that in the last decade big agribusiness – those who make up the top 10 per cent and are also the richest – cornered a healthy 72 per cent of government support.[5]

We see the perversity in advice given to African governments, which urges them not to subsidise the agricultural sector, when we consider that this is the sector that provides the bulk of employment on the continent and directly places food on the tables of the majority of the people. Place this reality side by side with the situation in the North where only a tiny fraction of the population is involved in agriculture. In terms of overall economic output, agriculture fills only a slim space. In France, for example, it contributes 2.9 per cent of gross domestic product (itself a biased measure); in Japan it is 1.6 per cent. Placed side by side with aid flows, agricultural subsidies in the North far outstrip the amounts given as aid to Africa.

With this contradiction in mind, the question that can logically be asked here is: does the North deliberately aim to hamper progress in Africa? Some analysts have responded by saying that this is not the case, but that as trade increases over the years, and due to rapid globalisation, there has been a need to develop some regulatory mechanisms to ensure that things are predictable. Unfortunately, they say, since the North is the dominant player in such negotiations, they ensure that the agreements and rules do not harm their own economies. And, unfortunately, they harm someone else. Some of these measures ensure that free market forces are only allowed to operate where it suits the interests of the dominant players. An example is the brain drain that has hit Africa over the years, with skilled African labour getting readily absorbed into the workforce of the North while less skilled folks are routinely refused entry into those same markets.

Let us look at the SAP cocktail and what it aimed to achieve. The selling pitch of the SAPs and current economic advice remains the claim that the strategies will place the continent on a new path to development through the export market. Progressive regimes in other continents have, in contrast, aimed to achieve import substitution, which would obviously lead to a leap forward for their countries.

Beginning in the 1980s, the IMF and World Bank imposed a neoliberal economic agenda on African countries seeking financial assistance including debt relief. This agenda is captured in what is generally referred to as conditionalities. They included:

- Minimisation of the state through privatisation of state enterprises
- Liberalisation of the economy to aid resource extraction as well as export-oriented open markets. This included the elimination of import controls
- Reduced protection of domestic industries
- Higher interest rates and austerity to lower consumer demand and thus curtail inflation, and encourage savings. Egged on by Washington, most African leaders demanded that citizens tighten their belts, while the leaders' waistlines widened
- Elimination of subsidies on food and agricultural products
- Removal or weakening of financial regulations such as currency controls, with the aim of attracting foreign investors.

On the whole, the SAPs were assembled to ensure that African countries would reduce their deficits on external accounts and at the same time achieve a balanced government budget. The conditions set by the IMF and the World Bank were bound to cause more problems than the ones they were supposedly set to solve. Cuts in public expenditure meant that social programmes were the first to be hit. Education, healthcare and housing readily received the hammer while spending on the military and police was typically raised to keep the unruly masses quiet.

The introduction of trade liberalisation meant the dumping of cheap agricultural and industrial products on the continent thereby undermining local production. Such imports included rice, wheat, milk, chicken and even maize. The immediate outcomes included the collapse of agriculture as well as fledgling industrial sectors. By 1989, Adebayo Adedeji and other economists at the United Nations Economic Commission for Africa proposed the 'African Alternative Framework to Structural Adjustment Programs for Socio-Economic Recovery and Transformation' as an alternative to the orthodox prescriptions of the World Bank and the IMF. When this framework was presented at the United Nations General Assembly in November 1989, it was welcomed as 'a basis for constructive dialogue'. It is very poignant that only one country voted against the resolution. That country was the United States.[6]

As described by one of the world's leading maverick intellectuals, Noam Chomsky of the Massachusetts Institute of Technology, Washington's ideology opposed Adedeji's ideas on principle:

A central part of neoliberal doctrine is what's called 'minimising the state.' That's long-standing World Bank doctrine. It runs exactly counter to its analysis and technical studies, and now even its proposals, which is maybe something worth thinking about. But putting that aside, minimising the state means maximising something else. That is, decision-making continues. It doesn't disappear, but it continues elsewhere. Where does it continue? The doctrine tells you that it shifts to the people, but again, that hardly rises to the level of a bad joke. The decisions are shifted into the hands of unaccountable private power, actually totalitarian institutions that are unaccountable, and it shifts away from the public arena, where at least in principle there can be some measure of public participation and influence and control. That amounts to a very sharp attack on democracy. It's intended that way and obviously has that character. And it's not at all surprising that private power appreciates it, and it's only reasonable to suppose that it will continue. At least they will try.[7]

The African framework recognised that the SAPs failed to address the need for improved social and technological infrastructure and failed to mobilise the enthusiasm, support and creative abilities of the people and grassroots organisations.[8] Moreover, they led to a loss of jobs, closure of businesses and thus social upheavals – 'IMF riots' – across the continent as basic necessities disappeared from market shelves or were out of reach of a seriously disempowered people.

This framework diagnosed the underlying causes of Africa's socio-economic crisis as the structure of an economy:

> that obliges Africa to keep producing commodities it does not need because its people consume very little of such commodities while it depends on other people for the production of its own needs. It is a structure of dependency rather than self-reliance. It is a structure that is more import-export oriented than production-oriented.

It went ahead to suggest that Africa must set up an African Economic Community by the year 2000 as envisaged by the Lagos Plan of Action. The African Economic Community would work to utilise the advantages of the continent and promote product specialisation for a large and unified African market. This would happen if trade barriers within the community are removed and if there is an agreement on steps to reduce competition between countries in the sub region, such as by rationalised product specialisation. The third proposed plank was that the community should pool resources for research and development and engage in freely sharing experiences in the application of research results. All of this embodied the spirit of democratic planning. Nothing could be further from Washington's agenda.

Joseph Stiglitz, former chief economist at the World Bank, was critical of the Bank and the IMF:

> The IMF likes to go about its business without outsiders asking too many questions. In theory, the fund supports democratic institutions in the nations it assists. In practice, it undermines the democratic process by imposing policies. Officially, of course, the IMF doesn't 'impose' anything. It 'negotiates' the conditions for receiving aid. But all the power in the

negotiations is on one side—the IMF's—and the fund rarely allows sufficient time for broad consensus-building or even widespread consultations with either parliaments or civil society. Sometimes the IMF dispenses with the pretence of openness altogether and negotiates secret covenants.[9]

In 1999 the IMF replaced structural adjustment programmes with the Poverty Reduction and Growth Facility (PRGF) and its Policy Framework Papers with Poverty Reduction Strategy Papers (PRSP) as the new preconditions for loan and debt relief. The PRSPs are nothing more than a repackaged SAP. At a meeting in May 2001 in Kampala, African civil society groups criticised PRSP as mere window-dressing to improve the IMF and World Bank's declining legitimacy. They also saw the content of PRSPs as putting corporate rights before social, human and environmental rights. The programme blocked any space for inputs by local people while granting the IMF and World Bank more control 'not only over financial and economic policies but over every aspect and detail of all our national policies and programmes'.[10]

The PRSPs were introduced ostensibly in recognition that nationally owned participatory poverty reduction strategies should provide the basis of all World Bank/IMF concessional lending and for debt relief under the enhanced heavily indebted poor countries (HIPC) initiative. The HIPC was launched by the G7 in 1996 at their summit in Lyon and was 'enhanced' at their 1999 summit in Cologne, supposedly to provide deeper, more rapid relief to a wider group of countries, and to increase the initiative's links with poverty reduction.[11] Three dozen countries are said to have benefited from the HIPC by the end of June 2009 while 26 others have reached the completion point. According to the World Bank, a country enters the race to be classified as a heavily indebted poor country when it can show that it has:

> a current track record of satisfactory performance under IMF and IDA-supported programs, a Poverty Reduction Strategy (PRS) in place, and debt burden indicators that are above the HIPC Initiative thresholds using the most recent data for the year immediately prior to the decision point. At the decision point, many creditors, such as the Bank, the IMF, multilateral development banks, and Paris Club bilateral creditors, begin to

provide debt relief, although many of these institutions maintain the right to revoke this if policy performance falters.[12]

The first steps include signing an agreement with the IMF for three years and within that time the country's economic decisions must be first approved and then strictly monitored by Washington. The conditions attached to the HIPC, as noted by civil society groups, present SAP in another garment. Macroeconomic policies such as the interest rate and currency (monetary policy), government budgets (fiscal policy) and what the country imports and exports (trade policy) are off the table, not up for negotiation with civil society. Some tinkering at the margin is permitted but most NGOs that participate in HIPC adopt the neoliberal premises so do not challenge power.

When the three-year period ends, the World Bank and the IMF conduct a review process to determine if the policies adopted by the country are sufficient to enable it to repay its debt. How do they figure this out? They check to see if the ratio between the present value of the debt the country owes and the annual value of export revenue exceeds 150 per cent. Anything above that is considered unsustainable and the country thus qualifies to be classed as a HIPC.[13]

The HIPC initiative currently identifies 40 countries, most of them in sub-Saharan Africa, as potentially eligible to receive debt relief (see Table 1).[14]

Did HIPC debt relief deliver the goods? Not according to the IMF, which admitted in its 2009 report 'The implications of the global financial crisis for low-income countries' that debt relief was only a matter of a reduced stock of debt, but the flows of debt payments actually stood to worsen. In other words, the Gleneagles debt relief promises were only designed to continue milking Africa.

Before we move from reviewing structural adjustment and its metamorphosis, we should briefly examine the World Bank and the IMF themselves. These two bodies claim to be inclusive institutions that respect the equality of nations as epitomised by the one-country, one-vote United Nations. The fact is that the IMF operates more like a transnational corporation and entry is dependent on the purchase of shares by governments. However, a country cannot just walk up and decide to buy up the majority of shares

Table 1 Countries identified as potentially eligible for HIPC debt relief

Completion point (26 countries)		Decision point (9 countries)	Pre-decision point (5 countries)
Benin	Malawi	Afghanistan	Togo
Bolivia	Mali	Chad	Comoros
Burkina Faso	Mauritania	Côte d'Ivoire	Eritrea
Burundi	Mozambique	Democratic	Kyrgyz Republic
Cameroon	Nicaragua	Republic of	Somalia
Central African	Niger	Congo	Sudan
Republic	Rwanda	Republic of	
Ethiopia	São Tomé and	Congo	
The Gambia	Príncipe	Guinea	
Ghana	Senegal	Guinea-Bissau	
Guyana	Sierra Leone	Liberia	
Haiti	Tanzania		
Honduras	Uganda		
Madagascar	Zambia		

in the IMF. The ratios are so well guarded that the controlling powers remain in the North. By its internal workings, the managing director of the IMF must always be produced by the Europeans, as witnessed in the farcical process of replacing Dominique Strauss-Kahn in 2011. The number two spot is reserved for the US. The World Bank is similarly apartheid-ordered, but instead of a 'Europeans Only' sign (as found in South Africa before 1994), the World Bank president's door says 'US citizens only'.

Can these institutions be reformed given these kinds of constraints to democracy? Strauss-Kahn proudly claimed that the question of IMF legitimacy was settled in November 2010, when he arranged that at some future date the European Union might give up seats and voting power to slightly balance the institutions, but it turned out that the main beneficiary of a token 6 per cent shift in voting power was China – no friend of the African poor and environment – and Africa's voting weight declined.

Africa has the smallest share of voting rights at the IMF, disadvantaging the continent's political capital in critical decision-making policies. The media publication Bloomberg, in an article entitled 'IMF electoral math doesn't add up', stated:

The distortions date back to the 1940s when the developed world of Europe and the U.S. had most of the world's cash, know-how and leadership. Voting rights were skewed accordingly. Under a gentleman's agreement among nations, the U.S. claimed the top job at the World Bank, which provides development aid, while Europe took the IMF, with its emphasis on government rescue loans and austerity plans. ...

To see what's wrong with the electoral math, consider Belgium and Brazil. Belgium is the world's 20th largest economy, with a 1.86 percent voting share in the IMF. Brazil is a vastly larger and more populous nation, ranked in the world's top 10 economies, with triple Belgium's output. At the IMF, however, Brazil is the weakling with just 1.72 percent of the vote.[15]

A further problem is ideology, as the case of Hernando de Soto shows.

The ideology of living capital

In order to lower political and legal risks, a fully fledged, pro-corporate rebooting of an African country's property rights can assist enormously. A pro-poor ideology comes in handy. According to Anuradha Mittal of the US-based Oakland Institute, in countries like Zambia, foreign investors can acquire land by offering the chief as little as a bottle of Jack Daniels whiskey. This is often portrayed as the product of a lack of property titling in Africa, where land is shared, often through customary laws. But this is messy, and a better arrangement is preferred. Across the world, such informally controlled assets are described by Peruvian economist Hernando de Soto as $9 trillion-worth of 'dead capital',[16] locked away from the billions of rural poor because of the lack of private property rights (through titling) and securitisation. Once a property right is applied to land or a house, then it can be registered and a lien taken by a bank in a deeds office, and this gives the owner the possibility of drawing credit – but at the risk of losing the collateral. With the collapse of much of the hype over microcredit because of disasters in South Asia (200,000 debt-related suicides in the Indian state of Andra Pradesh since 2000, for example) and, indeed, across the world, this is a dangerous recipe for a new round of displacement.

Yet, de Soto is a celebrity adviser to governments, ranging from Aristide's in Haiti prior to 2004, to the UK's Downing Street, to Washington's White House and Russia's Kremlin. In 2004, de Soto was even described by Bill Clinton as 'the world's most important living economist'.[17]

Ominously, however, de Soto has co-opted the language of justice by defining the premise of poverty as lack of access to private property and credit. From there comes many of the ideological pressures associated with privatisation.

Take Uganda: in 1998, President Yoweri Museveni – eager to portray himself as a new kind of African leader ushering in a dawn of transparent governance – began a land privatisation programme creating, in theory, a property market responding to market 'demand'. This rendered the buying and selling of land a far easier, and simpler, issue for investors. District land boards began the process of selling land. But for those rural people without political access, legal power and the finance to compete with high-powered interests, as *Le Monde Diplomatique* writes, 'land reform became a land grab.'[18]

So, who precisely does this legalism benefit?

In an interview with *World Policy Journal*, de Soto describes the process by which foreign investors receive inviolable property rights. These same rights, he claims, should be given to the world's poor:

> So here comes an American company, for example. They get a property right from the Peruvian government and that property right is in the form of a concession. It establishes that it's theirs, provided you renew agreements and commitments over so many years, but it's a property right. And they're going to have it for 60 years, 120 years, or it's unlimited ... Then that property right – which has been protected by the bilateral international treaty that has been signed between Peru and the United States – gets inscribed and blessed by the United States Overseas Private Investment Corporation (OPIC), which issues guarantees that makes it interesting for everybody to look at that property, because it's got the backing of OPIC. Then that property right goes to the Multilateral Investment Guarantee program of the World Bank, where again it is ratified by its 187 member countries, who of course view this as a way of guaranteeing investment in developing nations.[19]

According to de Soto, this 'perfected' property right can be possessed by the owner to any critical investor or financial hub, from Wall Street to London, as a title that, 'even the Peruvian congress can't take away'. But, stated de Soto, while tribes may have the same amount of land right next door, the difference – 'strictly the paper' – is both vast in practical terms and vastly unequal.

Yet land titling and all that accompanies it does not fall into a vacuum but into a specific socio-economic circumstance: neither the OPIC, World Bank, nor any other neoliberal institution will take the time or trouble to implement the same careful rights for the poor. As Ugandan historian Phares Mutibwa complained, the taking of Bugandan land by foreigners, 'is causing a lot of resentment. There is bad blood between the new owners of land in Buganda and the Baganda peasants.'[20] Unsurprisingly, perhaps, de Soto's book *The Mystery of Capital*[21] claims that the US's Wild West frontier example of property titling helps explain that country's success. But there is no mention of the loss of the native peoples, save that they did not understand the concept of 'owning' land.

Enforcing the ideology are the Bretton Woods institutions and their SAPs. As already described, the SAP may well be the single most potent tool used in the emasculation of the African continent, ensuring that the continent retains an empty bowl in her hands, ready to welcome further help that further deepens her vulnerability. The kinds of tensions generated in the process are conducive to the outbreak of Africa's resource wars, another case of the international system amplifying the problems caused by extractivism.

Resource wars

Central Africa has presented special difficulties when it comes to handling conflicts within the continent. The conflicts in Sierra Leone, Liberia and Côte d'Ivoire tangentially involved neighbouring states, but were nothing compared to the intricate entanglement of the states in the Central Africa wars. Conflicts that involved Rwanda, Uganda, Democratic Republic of Congo (DRC) and Angola sometimes got so complex that it was never clear whether combatants were proxies for silent parties.

In those conflicts and in the Somalia and even Sudanese debacles, the role of external advisers, notably the United Nations, has been called into question. One thing that has been clear is that underlying causes of the conflicts are often the struggle for the control of natural resources. While fighting in the region must involve mercenaries, as is to be expected, there are few cases in which external forces have openly intervened. The wars of independence of Angola and to a lesser extent Namibia were fought with the strong presence of Cubans.

The global divide erected by the cold war mostly determined the question of who was fighting on whose side in those years. If a side was supported by the Soviet Union, the other side was sure to obtain support from the US. The matter of what was legitimate was often pushed to the back burner. In this way dictators could be sustained in power no matter how evil they got as long as their presence in the seats of power secured the interests of the propping powers and blocked the access of those on the other side. This scenario worked well for patently corrupt leaders such as President Mobutu of Zaire (now the DRC).

With regard to the Rwandan genocide, it would appear that while Belgian divide-and-rule dates back generations, the immediate seeds of the conflict were sown as early as in 1990 when Tutsi exiles in Uganda formed the Rwandan Patriotic Front and aimed to seize power in Rwanda. The ensuing conflict attracted attempts at settlement that included the Arusha Peace Agreement of 1993. However, the assassinations of the presidents of Rwanda and Burundi in a 1994 plane crash led to a power struggle that triggered the genocide, leaving a million victims and about three million refugees out of a population of 7.5 million. At the time of the massacre, a United Nations Assistance Mission for Rwanda was already in place. It could not stop the bloodshed, in part because, as US President Bill Clinton later admitted, the people of Rwanda were simply not priorities in 1994.

Are ethnic differences always a major cause of conflicts in Africa? In the Rwandan case, the Hutus and the Tutsis are known to widely intermarry. The ordinary people lived at peace with their neighbours in the true African tradition. The problem is with the leaders who seek to create enclaves and territories for themselves; they find the ethnic card easy to play. They use these cards to sow suspicion, hatred and fear in the hearts of their people and literally lead them to slaughter their neighbours. As it is with the Hutus and Tutsis of Rwanda, so it used to be with the Ijaws, Itsekiris and Urhobos of the Niger Delta of Nigeria.

Conflicts in this part of Africa form a violent brew: ethnic cleavages, political power and control of resources. As much as outsiders intervene to join in the exploitation of the natural resources in the region, blame for this state of affairs must rest squarely on regional leaders. Take the case of the war waged by the late Laurent-Desire Kabila and his Alliance of Democratic Forces for the Liberation of Congo-Zaire (ADFL) against the Mobutu government in 1996. The ADFL was supported by Uganda, Rwanda, Burundi and Angola. The Angolans supported them allegedly as a payback for Mobutu's support for Savimbi's UNITA against the Angolan government. Kabila began to work against his allies after a year in office and that raised new levels of conflict as he aligned himself with the Hutus in the region. After the assassination of Kabila, in January 2001, his 29-year-old son took over the reigns of power.

At the height of the conflicts in the Congo axis there were up to nine armies in the fray. The diamond market that thrived in Kigali, even though Rwanda does not have any diamonds, highlighted the plunder of DRC's resources. A United Nations report issued in April 2001 accused Uganda and Rwanda of plundering diamonds, copper, cobalt, gold and coltan from DRC. Zimbabwean and Angolan commanders who helped Kabila even enjoyed special spoils of war: they were given rights to mine precious minerals in the zones that they defended.[22]

In many of these conflicts, the United Nations forces have played a peacekeeping role and are usually not involved in direct combat except in self-defence. However, there have been accusations that the peacekeepers fighting on the side of DRC government forces attempted to push the militia of the Democratic Forces for the Liberation of Rwanda (FDLR) out of DRC and into Rwanda. The United Nations denied complicity in the atrocities that followed the attacks. The United Nations Mission in Congo (MUNUC) is said to be the world's largest UN peace-keeping mission with 20,000 peacekeepers and a budget of $1.3bn in 2009.[23] Generally, Africa has attracted larger number of UN

73

peacekeepers than any other region. In the 1990s, out of 32 operations, 13 were in Africa.

Nevertheless, the UN failed spectacularly in Somalia and this appears to have put a damper on the outlay of resources and interventions elsewhere. Analysts blame the United States for the failure of the Somali state, given its role in encouraging Ethiopian president, Meles Zenawi, to invade in early 2007. The US still has deep interests in how the situation plays out in that country as they back the unity government that somehow oversees the nation. During her visit to Africa in 2009, the US secretary of state, Hillary Clinton, claimed that Eritrea was supporting Somalia's al-Shabab militants, who are fighting to topple the unity government.[24]

These advisers – including corporations that purchase great swathes of land (such as Jarch, as we discuss below) – seem to have a predilection for working primarily for their own self-interest rather than the interests of the nations they claim to be helping. Africa has been a ready space for scenario planners to test out their dreams. There are reports of mining companies sponsoring rebel groups in countries such as Sierra Leone so as to secure the space for the companies to loot the region of mineral resources in exchange for cash with which the rebel leaders can procure weapons and live flamboyant lifestyles. Countries with resource-rich regions sitting on their borders have used the same tactics. Both Uganda and Rwanda have been accused of sponsoring cross-border conflicts in order to grab mineral resources in DRC.

Liberia suffered one of the worst civil wars. For many people, the story of Liberia is one of blood diamonds. For others it is a story of rubber – with Firestone at the heart of the problem. But there is another sector that must not escape our focus in this country and that is the iron and steel sector, led by Lamco, a Liberian-Swedish-American company. It closed shop in 1989, in the wake of the country's first civil war. ArcelorMittal, currently the world's largest steel entity, entered the scene in 2005, negotiating a 25-year concession to mine and develop iron ore around the country's north-west border with Guinea. After two decades of vicious civil war, Liberians were eager to get the economy back up and running. Arcelor-Mittal engaged with the then-National Transitional Government, just three months prior to democratic elections, and moved for the juiciest agreement they could get: the bargain bin price of $900mn. That deal gave them astonishing concessions, including control over how much royalty they would pay, because, according to Friends of the Earth, the mineral development agreement does not 'specify the mechanism to set the price of ore and leaves open the basis for intra-company pricing, creating a strong incentive for Mittal to sell the ore at below market value to an affiliate, which would reduce the actual royalties paid'.[25]

This is not uncommon. Corporate mispricing facilitates over 60 per cent of Africa's annual illicit flight, estimated at a minimum of $148bn annually. ArcelorMittal not only created a company structure that shielded the parent company from being held responsible for the actions of the subsidiary. The company was also given the exclusive right to two major Liberian national infrastructures: a railway line and the Port of Buchanan. The only time the government of Liberia could use the port or the railway was if there existed spare capacity, effectively granting the company quasi-governmental status. Among other heavily manipulated clauses, ArcelorMittal was allowed to maintain its own unchecked security force, was exempted from human rights or environmental legislative acts introduced in Liberia, and was given the right to expropriate land almost at will.[26] Indeed, in a country emerging from a deadly civil war, this was a blank cheque for gross violation of both first generation (civil and political) as well as second-generation (economic, social and cultural) rights.

We know of these problems thanks to courageous non-profit organisations such as Global Witness, authors of the 'Heavy Mittal' report,[27] which was used by the government of President Ellen Johnson-Sirleaf as a guide to contract renegotiation. She attempted to junk some of the more exploitative clauses, including the use of the railway and the port, and the tax holiday. After intense negotiations, the new agreement was signed in December 2006 and ratified by the Liberian legislature in May 2007.

But notwithstanding the improvements, the environmental and social impacts remain contested. The concession area lies in the Nimba Mountain on the boundary with Guinea and Côte d'Ivoire. The sides of the Nimba Mountain in the neighbouring countries have been a nature reserve since 1944 and currently an area amounting to 180sq km has become a nature reserve as well as being named a World Heritage Site. As a nature reserve, the area is strictly protected and is not open to tourism. The Liberian side of the mountain is curiously not a protected zone. Is iron stronger than nature and life? The loss of biodiversity due to mining activities in this area is an open secret, but no one is willing to plug the hole.

Given the sky-high rate of unemployment, Liberians who are able to find work with the iron company are not in a good position to negotiate fair wages. According to Friends of the Earth, they 'work six to seven days a week in the tropical heat, and sleep in tents on a plywood floor. They are allotted a cup of rice and soup daily, drink well water and have their pay docked for hospital visits.'[28]

Jarch takes the Southern Sudan

Another revealing case of destructive extraction is Southern Sudan, where an American company banked on the break-up of the nation as the best outcome for their business interests. What makes this case more interesting is the suspected link between the corporation involved and its home country's State Department – and the possibility that this is a front for a foreign policy game. While Jarch Capital took a determined stand in Southern Sudan, Chinese companies were present in the north and continued to dig deeper.

Though it was illegal to do business in Sudan, Jarch was active long before the country's 2011 split. The company grabbed a 50-year lease on 400,000 hectares of farmland in Southern Sudan's Mayo county from Matip Nhial, the deputy commander-in-chief of the Sudan Peoples Liberation Army. In addition, Jarch Capital holds a 70 per cent stake in another company with rights to grow crops and process them into finished products for local as well as export markets. The company's motto, proudly displayed on its homepage in 2011, is 'Because It Is YOUR Land, YOUR Natural Resources!'

The firm's chairman and chief executive officer, Philippe Heilberg, told *Oil Review Africa*:

> We are a rather unique company in a unique niche in that our company focuses on countries in Africa that are undergoing and may undergo sovereignty changes such as changes to

international borders and the creation of new countries out of current ones ... we attempt to become experts on the geo-politics of those countries undergoing sovereignty changes and establish key relationships with the leadership or potential leaders of the new state. In addition, we attempt to sign natural resources agreements with these leaders or potential leaders even before they become recognised sovereigns ... [Jarch] believes in the empowerment of the populations who actually own the resources, sometimes being exploited by others. Jarch looks to work with the population to develop strategies to secure their political and economic rights of self-determination.[29]

According to researcher Sam Urquhart:

[T]he firm boasts a curious cast of ex-intelligence officials, Clinton era insiders and veteran neocon warriors on its management team. There's Larry Johnson (ex-CIA and frequent commentator on cable news about terrorism related themes). There's Gwyneth Todd, an ex-Clinton adviser on Middle East affairs. There's even Joseph Wilson, the ex-U.S. ambassador to Niger who became notoriously embroiled in the preparations for the War in Iraq. Alongside Wilson, there's also J Peter Pham, a prominent neo-con commentator, particularly on African issues. And, at the apex of the operation, there is Philippe Heilberg. Heilberg himself, frequently described as a bucaneering billionaire, has done time at AIG and Citibank (a glorious resume for sure)...[30]

This company's game plan in South Sudan falls squarely into the niche carved for corporate conflict entrepreneurs who seek out countries in socio-political turmoil and then pitch their tents. As the company's website brags, 'Since 2002, Jarch has focused primarily on Africa as the world leader in the growth potential for natural resources development.'[31] South Sudan became an independent state in July 2011, and Jarch is a case in which corporate interests benefit from stoking the fires of conflict in African countries, where self-interest entails splintering states.

But Jarch is not finished, for as Urquhart explains:

Cecil Rhodes would have been truly proud. In fact, Rhodes' stratagem of raising paramilitary forces to break up the Boer state in Transvaal anticipates Jarch's own by a century or more. It's classically imperialist. In fact, Heilberg has undertaken a fairly idiosyncratic reading of imperial history to prepare himself for his African adventures. As he told reporter Barry Morgan in February 2008, 'I look at the unwinding of the Ottoman Empire, the collapse of the Soviet Union, the two Gulf Wars and the fall-out from Bosnia & Herzegovina and I can see things reverting to a natural state of equilibrium where it simply makes no sense for some nations to be grouped together or stay in their earlier alignment.' A year later, he was telling reporters 'several African states, Sudan included, but possibly also Nigeria, Ethiopia and Somalia, are likely to break apart in the next few years, and that the political and legal risks he is taking will be amply rewarded.'[32]

The intractable nature of these conflicts can also be found in the fact that the same resources that ignite the conflicts also keep the flames burning, because once any of the contending parties lay their fingers on the resources they then use the resources to fuel their war machines and pound their war drums. The additional sad note is that sometimes the combatants forget why they started the conflict in the first place and, even when the original causes have gone, the fight goes on anyway.

Destructive extraction

Popping
A million explosions
A shower of soot
On open raw nerves
Oil's not well
That starts a well

Now the earth is ablaze
Where will the people go?[1]

THE EXTRACTIVE INDUSTRIES would like to have everyone believe that they operate a level playing field whether they are working in North America, Europe, South America, Asia or Africa. When prominent cases to the contrary are highlighted, some extractive sector transnational companies are quick to say that they are mere scapegoats and that they are not the worst corporations in the field. Some industry players have told me to my face that when we complain about oil corporations' behaviour in Nigeria we are merely biting the fingers that feed us. Some have gone so far as to say that without the operators Nigeria would starve and would be nothing, merely Africa's paper giant.

Shell, the giant oil corporation, has been much condemned for its atrocious record in Nigeria – and it constantly wriggles and screams about it. Its major problem is that no matter the denials, the facts unfortunately speak clearly. We can argue that the corporation has not always set out to do evil, and that the fact that it inevitably did so was a matter over which it had no choice.

We can argue that way. Yet that would not equate the penance of the penitent. The company could claim that it is a product of the environment in which it operates. And we would agree that the Nigerian environment is peculiar in many ways. But no one can claim absolution by heaping the blame on the economic or political environment. As Human Rights Watch put it, 'Companies have a duty to avoid both complicity and advantage from human rights abuses.'[2]

Documenting extractive destruction

Opting out of Nigeria, or Africa as whole, is a threat that corporations make but are unlikely to follow through, given the very high profit rates. In Nigeria, Shell officials have argued that it is not fair for them to receive such a heavy bashing for the social and environmental problems of the Niger Delta. At one point they even pointed fingers at their competitor, Mobil, saying that since it operates mainly from offshore fields it seems that they are excused because it is a case of 'out of sight and out of mind.' They also claimed that Nigeria would be worse off if Shell were to depart, as no other corporation that could take over their operations could boast a better record in health, safety and environmental standards.[3] A major study[4] carried out by Professor Richard Steiner reached the conclusion that Shell Nigeria operates well below internationally recognised standards to prevent and control oil spills. He gave the following reasons to back this conclusion:

- Lack of implementing 'good oil field practise' with regard to pipeline integrity management (particularly the US IM regulations, API standards, and Alaska's best available technology requirements)
- Delay in initiating an asset integrity review (AIR) and pipeline integrity management system (PIMS) for Shell Nigeria, which admits it has a backlog in its asset integrity programme
- Questionable adequacy of Shell Nigeria's AIR and PIMS, and lack of independent oversight
- Lack of reference to and attention by Shell Nigeria to the Niger Delta as a High Consequence Area for oil spills

- Lack of adequate attention by Shell Nigeria to the Niger Delta as an area in which oil facilities are susceptible to intentional third-party damage, requiring enhanced pipeline integrity and monitoring procedures
- Exceptionally high number, extent, and severity of oil pipeline spills in the Niger Delta before, during, and after their AIR and PIMS
- Lack of transparency in Shell Nigeria – the AIR, PIMS, joint operating agreement and its oil spill contingency plan should admit to independent third-party evaluation
- Lack of adequate oil spill response capability and performance of Shell Nigeria.

The Steiner report focused on oil spills, but it gives us a picture of the general situation in the oilfields. Bopp van Dessel, a former Shell head of environmental studies in Nigeria, quit the corporation in 1994 citing professional frustration following Shell's extensive pollution and breaching of international standards. He was quoted as saying: 'They were not meeting their own standards, they were not meeting international standards. Any Shell site that I saw was polluted. Any terminal I saw was polluted.'[5] To be fair, Shell itself has admitted that its environmental standards are lower in Nigeria than in America and Europe.[6] The corporation also admitted to Christian Aid that the overall picture of the age and integrity of their pipelines in the Niger Delta as well as their (Shell's) transparency when compared to industry standards was incomplete.[7] Table 2 indicates the rate of pipeline failures in selected regions. It shows clearly that the corporation's operation in Nigeria is in a class by itself; certainly not aligned to international standards.

The number of oil spills and the regularity of their occurrence make nonsense of any claims to acceptable standards by any of the corporations operating in the Niger Delta. Although oil spill records diverge depending on the source, they all show that the volumes of toxins released into the environment are incredibly high. Combined, they make the world-famous Exxon Valdez oil spill of 1989 pale in comparison. That spill affected 2,000km of shoreline with 320km being heavily or moderately affected.

Table 2 Comparison of worldwide pipeline failure rates[8]

Region	Product	Failure rate per 1,000km-years	Year
United States	Gas	1.18	1984–92
United States	Oil	0.56–1.33	1984–92
Europe	Gas	1.85	1984–92
Europe	Oil	0.83	1984–92
Western Europe	Oil	0.43	1991–95
Western Europe	Gas	0.48	1971–97
Canada	Oil and gas	0.35	N/A
Hungary	Oil and gas	4.03	N/A
Nigeria	Oil	6.40	1976–95

Twenty years later, traces of the spill can still be found along the shoreline.[9]

Renowned Nigerian environmental law professor Margaret Okorodudu-Fubara estimated that 2,105,993 barrels of crude oil were spilled into the Niger Delta environment between 1976 and 1990 in a total of 2,800 incidents.[10] Outstanding incidents include the rupturing at Shell's Forcados terminal in 1979 where 570,000 barrels were spewed into the estuary and adjoining creeks.

Texaco had its day in 1980 at Funiwa, where 400,000 barrels of crude oil were emptied into coastal waters and destroyed 340 hectares of mangrove forests. For Mobil, their landmark spill was recorded in January 1998 when their Idoho platform released 40,000 barrels of crude onto the Atlantic coast, affecting at least 22 coastal communities.[11] In the case that ensued, Texaco had to reach out-of-court settlements with the affected communities.

In October 2009, the Nigerian National Oil Spill Detection and Response Agency (NOSDRA)[12] announced that 2,122 oil spill incidents were recorded between 2006 and 2009. The crude loss was put at 66,696 barrels. There were 252 incidents of spill in 2006; 597 incidents in 2007; 927 cases in 2008 while between January and June 2009, 346 cases were recorded.[13]

Table 3 Oil spills recorded in the Niger Delta over 23 years
(1976–1998)[14]

Year	Number of spills	Volume in barrels of oil
1976	128	26,157
1977	104	32,879
1978	154	489,295
1979	157	694,117
1980	241	600,511
1981	238	42,723
1982	257	42,841
1983	173	48,351
1984	151	40,209
1985	187	11,877
1986	155	12,905
1987	129	31,866
1988	208	9,172
1989	195	7,628
1990	160	14,941
1991	201	106,828
1992	367	51,132
1993	428	9,752
1994	515	30,283
1995	417	63,677
1996	430	46,353
1997	339	59,272
1998	390	98,272
Totals	**5,724**	**2,571,114**

The Niger Delta region of Nigeria has the reputation of being one of the most polluted places on earth. That reputation has been taken as a given, due to physical evidence of degradation. Possibly the most comprehensive environmental audit of the region is the Niger Delta Environmental Survey (NDES) commissioned by Shell Production Development Company (SPDC or Shell) in the 1990s. The results of that survey have not been made public as Shell chose to lock the reports up in its vaults.

However, after the federal government of Nigeria commissioned the United Nations Environment Programme (UNEP) to assess the environment of Ogoniland, a report of a segment of the Niger Delta is finally available. The assessment took 14 months to complete and the report was delivered to President Goodluck Jonathan of Nigeria on 4 August 2011. Among the key findings are:

> In at least 10 Ogoni communities where drinking water is contaminated with high levels of hydrocarbons, public health is seriously threatened ... In one community, at Nisisioken Ogale, in western Ogoniland, families are drinking water from wells that are contaminated with benzene – a known carcinogen – at levels over 900 times above World Health Organisation guidelines. The site is close to a Nigerian National Petroleum Company pipeline. UNEP scientists found an 8 cm layer of refined oil floating on the groundwater which serves the wells. This was reportedly linked to an oil spill which occurred more than six years ago.
>
> While the report provides clear operational recommendations for addressing the widespread oil pollution across Ogoniland, UNEP recommends that the contamination in Nisisioken Ogale warrants emergency action ahead of all other remediation efforts.
>
> While some on-the-ground results could be immediate, overall the report estimates that countering and cleaning up the pollution and catalyzing a sustainable recovery of Ogoniland could take 25 to 30 years.[15]

Although the report was not conclusive on health and crop yield impacts, the facts it exposed were sufficiently alarming. The researchers found, for example, hydrocarbon pollution up to a depth of 5m in the soils of Ogoniland.

Reading the UNEP report, knowing that oil exploitation was

halted in Ogoniland in 1993, it is easy to deduce that the other communities in the Niger Delta territory are as badly damaged or even more so, considering that new pollution events are still occurring, including the release of toxic elements from the ubiquitous gas flares.

Other responses to the UNEP report question some of its blind spots. For example, Professor Richard Steiner noted that:

> The UNEP report devotes several pages (161–166) specifically to artisanal refining at the Bodo West oilfield, and correctly reports an unfortunate increase in such between 2007 and 2011. However, in this analysis of oil pollution in this region, UNEP entirely ignores the other much larger source of oil spilled into this same region in that same time period – the twin ruptures of the Trans Niger Pipeline (TNP) caused by SPDC negligence in 2008 and 2009. Together these spills contributed between 250,000–350,000 barrels of oil into this system, orders of magnitude more than illegal refining. Much of the oil at Bodo West area likely derived from the TNP Bodo spills.[16]

The heart of darkness

Moving to the Congo basin, the mineral wealth has for generations attracted the attention of imperial powers. The rich deposits found here helped to fatten the coded accounts of despots and also helped to fuel corruption and violence in vicious ways. In the midst of the political challenges of the region, venture capitalists have learned to detect opportunity in chaos, which is obviously easy to do if you are a part of those orchestrating the chaos. The region remains attractive to mining companies because they can extract using the cheapest methods possible: open cast extraction and the backs of a compliant labour force oppressed into submission by years of tyranny. President Mobutu is said to have personally owned one of the Haut-Zaire gold mines while he also kept the operations of the Société Minière under his gaze.

The Katanga region provided a strong entry point for Union Minière (UM), which has been characterised as 'the epitome of colonialism' in the 1930s. By the time of independence, Belgian protection of its colonies had bequeathed Moise Tshombe a system capable of undermining democracy. Although UM had quit the country by 1967, they secured a strong stake in the Congolese Company GEOCOMIN and retained lucrative rights in the venture well into the 1980s. As icing on the cake, the chairman of UM, Edward Sengier, is said to have appropriated uranium from the Shinkabalowe mine and shipped it to the US in 1940. It is believed that Congo's uranium had its final resting place in the tragedy of Hiroshima and Nagasaki, five years later.[17]

Despite years of plunder, the Democratic Republic of Congo (DRC) still has a huge reserve of copper and is a very important supplier of cobalt. The country also has good deposits of gold, although most of the mining here is done by artisans who pine away as they pan for the yellow stones.

New mining codes allow generous incentives, which often include lowered taxes and royalty rates. In addition, conditions are made difficult for local companies to compete, suggesting all these tactics and more are aimed at attracting direct foreign investment.

Zimbabwe – unearthed

Although mining companies do not appear to mind mining in conflict zones, it does appear that at times they are wary of being associated with certain political figures. This appears to have been the predicament of Rio Tinto plc when they could not pinpoint exactly where in Africa their Murowa diamond mines are located. The mines are indeed in Zimbabwe, but due to the heavily tainted human rights record of President Mugabe, his unique economic policies and not so unique ways of winning elections, Rio might understandably want to blur things a little bit. A press statement issued by the company on 29 May 2008 gave general information such as that their Argyle diamond mine is in Australia, that the Diavik mine is in Canada, and that Murowa is simply 'in Africa'.[18] Could they have forgotten where the Murowa mine, a key contributor to their $1bn diamond business, is located?

The story of Zimbabwe and her diamonds would make your skin crawl. Private security forces as well as the public ones treat poor Zimbabweans who happen to work around the mines – such as the one discovered in Marange in 2006, which is estimated to hold $800bn worth of diamonds[19]– in a brutally exploitative manner. More recently, it was revealed by the BBC *Panaroma* programme that a torture camp known as 'Diamond Base'– a remote collection of tents closed in by barbed wire fencing – imprisoned miners who were subsequently tortured. According to one released prisoner quoted anonymously by the BBC, 'They beat us 40 whips in the morning, 40 in the afternoon and 40 in the evening.' The man still could not use one of his arms after the beatings and could barely walk. 'They used logs to beat me here, under my feet, as I lay on the ground. They also used stones to beat my ankles.'[20]

Zimbabwe has had its fair share of political challenges and constructs with many actors and players. The power-sharing arrangement between the Movement for Democratic Change (MDC) – Zimbabwe's main opposition party – and Mugabe's ZANU-PF has not done much to insulate the nation from the claws of destructive extraction. While land remains a contentious issue in Zimbabwe, the stepping up of diamond exploitation especially by Chinese interests has brought in new pains.

One of the most notorious of the Chinese players in the diamond fields of Marange, Zimbabwe is the China International Fund (CIF) – a private Chinese entity with the record of being the country's largest known foreign investor.[21] CIF has a convenient base in Hong Kong where the secrecy jurisdiction allows the thriving of a huge network of more than 30 opaque subsidiaries, known as the '88 Queensway Group'.[22]

Though Hong Kong does not provide banking secrecy, companies within its jurisdiction are not required to include details of trusts, company ownership, or beneficial ownership on public and official records.[23] There are speculations that CIF is closely tied to the Chinese state through revolving door officials as well as by backdoor financing, resource trading and hidden beneficiaries. Chinese officials refute these suspicions and insist that there are no official ties with CIF.

The company's ties with another African state, Angola, are not so easy to deny. It is said that leading officials of CIF are intertwined with Angolan officials such as Manuel Vicente, chairman of the board, as well as a senior person in the Angolan state oil company, Sonangol, through at least nine different joint ventures under the umbrella of China-Sonangol. Vicente is also known to run Angola's largest phone operator, Unitel, with the president's daughter, Isabel. In addition, it is believed that CIF is the source of $2.9bn in funding for Angolan construction projects, administered by Angola's Gabinete de Reconstrução Nacional, according to information given by a former top government official. Angola's lifetime dictator, Eduardo Dos Santos, counts Xu Jinghua, head of the CIF and also alleged to be a Zimbabwean arms dealer,[24] as an old military academy classmate in the Soviet Union.[25]

Interestingly, China-Sonangol and CIF are said to have invested $7bn in Guinea and $8bn in Zimbabwe.[26] The company, Sino-Zimbabwe, incorporated as Sino-Zim Diamond Limited with roots in Hong Kong, shows the close ties between international diamond magnate Lev Leviev, the Angolan government and the CIF.

Anjin is another company mining diamonds in Zimbabwe. The company, a joint venture, began operations in 2009 and held 10,000 hectares of land at Chirasika in Marange. The company later discovered new diamond deposits in Chiadzwa at a place local panners call Jesi. The company's land holding expanded rapidly to over 30,000 hectares over a short span of time. Moreover, the company operated behind a thick military shield. Indeed, locals suspect that some of the Chinese workers in the mines are actually military personnel.

What is not in doubt is that senior managers at Anjin run the mines like a military formation. Perhaps to ease relations between the Zimbabweans and the Chinese the company operates a dual form of administration with the Chinese running their own affairs in terms of staff management, while the Zimbabwean military take care of security-related issues.

The lack of transparency in the operations in the diamond fields is captured by reports stating that:

> In January 2011 the president addressed the chiefs' conference in Kariba and briefed them on the situation in Marange, high-lighting that the company had not started mining diamonds at Chiadzwa but was busy building houses for resettling people at Arda Transau in Odzi. One month later – February – the government announced that the company had a stockpile of one million carats and was awaiting KP [Kimberly Process] certification.[27]

In a situation like this, it is not at all surprising that the volume of diamonds mined here can only be speculated on.

Zimbabwe's Farai Maguwu, an imprisoned diamond researcher and human rights activist, who according to *The Economist* magazine was a 'first class' source, directs the Marange-based Centre for Research and Development (CRD). He says:

> Whilst I can't commit myself to mentioning names, our observations indicate that some very senior military personnel and well placed politicians are directly involved in the mining operations of Anjin. The involvement of the army in diamond mining in Marange is the saddest thing that has happened to the find of the century.[28]

Maguwu was arrested as an enemy of the state in 2010 because he was allegedly 'endangering national security' for having information relating to the Zimbabwean military's gross human rights violations at Marange's diamond mines.[29] Maguwu was arrested just as the appointed monitor of the Kimberley Process, Abbey Chikane, was to meet him. He simply walked into the embrace of state intelligence officials who were also present. Chikane claimed that failing to inform the Zimbabwean state would have been illegal. Charges brought against Maguwu were eventually dropped after considerable international lobbying by major local and international NGOs and civil society movements including Global Witness.

South Africa's eGoli

Historically, South Africa was said to have streets of gold. In fact, Johannesburg is even labelled as 'eGoli', the city of gold. This is both because the province of Gauteng, where Johannesburg is based, contributes over 30 per cent of GDP to South Africa's economy and also because the city itself has yielded some 45,000 tonnes of gold – first to colonialists like Cecil Rhodes, exploiting resources in the name of the British crown, thereafter through multinationals such as Anglo American, backed by the apartheid government, and more distantly, the US and UK, their partners in exploitation.

Now, when commercial publications like *The Africa Report* carry articles headed, 'Companies profit from toxic water',[30] it is time to sit up and take notice. An article published in the magazine claimed that beyond physical exploitation of resources lies a more devastating resource curse: environmental pollution, chiefly acid mine drainage (AMD). Characterised by high acidity, heavy metals, salts and sulphates, 40 million litres of high corrosive, radioactive and toxic AMD has leached into the local ecosystem, contaminating groundwater. The article claimed that in 2002, the West Rand basin began decanting from underground mines, spilling both over-ground and underground into interconnected tunnels and mining cavities. Though the projection was made in 1996, government inaction enabled mining companies – using just 7 per cent of water, but causing 75 per cent of pollution to South Africa's water-scarce environment – to elide the costs of pollution, classified as 'externalities'.[31]

Often, civil society movements and NGOs are brought in to represent the realities and 'interests' of 'external stakeholders'. But this time, so obvious was the case, that even people like hydrologist Garfield Krige, a former water technologist for mining company JCI who formed part of the environmental team that conducted the investigation, openly stated on the record that, 'Apart from having meeting after meeting and discussing the issues around the AMD, nothing in reality has been done since 1996.'[32]

He went on to reveal that governments' behaviour potentially aided mining houses to either dump mines or sell/transfer assets and liabilities (including AMD) to new owners. Active companies like DRD Gold justified the stance, claiming they were not responsible for the historical problem of pollution as they occupied the mine for just 5 per cent of its operational life and mined 0.5 per cent of the total number of tonnage exploited.

Even Mike Muller, a former director general of the Department of Water Affairs and Forestry, told the magazine that DRD's submission to the Security Exchange Commission in New York revealed 'substantial financial liabilities and warned that environmental issues related to water management were perhaps the single biggest liability facing the company'. DRD Gold recently declared profits of R100mn, chiefly attributed to the sale of old mines – which included environmental liabilities.

How much is the total price tag of water pollution calculated thus far? Africa Earth Observatory Network (AEON) report

'H_2O-CO_2 energy equations for SA' estimated, 'that water treatment and desalinisation plants to handle AMD would cost R360bn over 15 years.'[33]

When the government of South Africa's Department of Water Affairs (DWA) released the results of an expert report assessing the threat posed by AMD, the words used were, 'a matter of urgency'. The government proceeded to allocate public funds for the pumping and treatment of mine water in the western, central and eastern basins. But many of the corporations in question, past and present, live in denial. The agricultural trade association of South Africa, known as AGRI-SA, representing more than 70,000 commercial farmers, publicly declared: 'We know that South Africa is a water-scarce land and if we allow for contamination of our groundwater resources, we are heading for serious trouble.'

Obviously, the extractive sector machinery cares most about profits and while they pile up tonnes of 'precious' metals, the consequences of their activities receive precious little attention.

Iron-clad love

A while before Lansana Conte breathed his last as president of Guinea, he sent a distinctly unfriendly – and unexpected – letter revoking the licence of mining company Rio Tinto's interest in the Simandou mountain range's iron-rich mines. The project had the full backing of the International Finance Corporation (IFC)[34] – the World Bank's private-sector lending arm – which had invested $35mn in the iron mine venture. It was also not an easy season for the IFC, which at the same time was exploring the possibility of a $500mn stake in a massive aluminium project in the country. The close partnership or wedlock between transnational corporations and international finance institutions could not be better illustrated than here, where in the face of uncertainty, Rio Tinto could speak not only for itself but also for the IFC. Before the IFC could say anything, Rio Tinto stated that it and the IFC 'remain committed to the project'.[35] With a value put at $6bn it is easy to see why the IFC would not hesitate to climb into bed with the mining mogul. While this was going on, Chinese investors were already peeping through a crack in the door. The Guinean officials overseeing these deals could not sleep with both eyes closed because within the space of a week, Conte fired the official who first raised questions on the Simandou agreement, promoted him, and then sacked him again.

Interestingly, an official of Transparency International (founded by a former World Bank official) sided with the mining company, claiming that Guinea 'would miss out on all the benefits' and that 'the big losers would be Guinea's impoverished people if the project was cancelled'. It was not clear how much Guinea's poor would stand to gain from the project, and what all of the benefits would be. At that time, Transparency International ranked Guinea as one of the world's most corrupt countries. And yet this official believed the same system would deliver public goods?

With the entry of the military junta into the power equation in Guinea, Rio Tinto may have thought that things would become rosy once more. They were in for a rude shock when the military junta ceded half of Rio Tinto's acreage to a rival mining company, BSG Resources, owned by Israeli billionaire Beny Steinmetz.[36] The battle for control of the mountain range mining area raged. Rio Tinto insisted that it retained full mining rights to the acreage but the Guinean government called their bluff and declared flatly that its decision was final. While the company claimed that they need the entire acreage to make the project viable, observers believed that their grouse was that the half that was given to BSG Resources was richer in ore than the section they held.

In the struggle to retain mining rights over the entire area, Rio Tinto also used the local poor communities to plead with the government on the company's behalf, on the grounds that some of the villagers were employed by the company and might lose their jobs if the company did not retain the entire area. In other public advocacy, Rio Tinto claimed that their concession was only 738sq km while 3,333sq km was held by another unnamed company. They also claimed that Rio Tinto's concession area represented only a fraction of the over 20,000sq km of iron ore exploration title awarded to other companies in Guinea.[37]

In March 2011, the *Financial Times* lauded the rule of Guinea's new president, Alpha Conde, who had ended two decades of brutal dictatorship and corruption.[38] Conde's supporters include powerful backers such as George Soros, described as his 'billionaire philanthropist adviser', working to help devise new mining codes. Others, like Mamadou Sylla, one of Guinea's richest men, a renowned arms dealer and former cheerleader for Conte, have less reputable interests.

Corruption is big business in Guinea: the 2008 military coup under Captain Moussa Dadis Camara facilitated the public sector making 85 procurement deals worth $1.5bn.[39] Just three followed the legal tender procedures. Until the coup, Camara, friendly with Conte, was barely known outside particular military units. His reasons for the coup? 'I had no choice after the death of the president: I had to either take over or leave the country … had [then head of the Guinean armed forces Diarra] Camara taken over, I would have to go into forced exile to escape a certain death!'[40]

When looking at the broad strokes, Guinean President Alpha Conde, the country's first democratically elected leader, seemed to make a strong move toward justice. But the finer details reveal otherwise: over 42 per cent of all the public tenders from the 85 deals were allocated to Kerfalla Person Camara, head of building corporation Guicopress.[41] Conde's Rassemblement du Peuple de Guinée (RPG) party was largely financed by Kerfalla.[42]

In July 2011, Conde was almost assassinated when his residence was attacked. He expressed words of gratitude to the presidential guard who valiantly defended his compound.[43] 'Our enemies will not be able to stop Guinea's progress,' said Conde, who is credited with ending two years of military junta rule. Yet, according to one *Africa News* reporter based in Conakry, 'The demonstration of opposition parties in Guinea had been violently repressed by security forces in Conakry. On Tuesday, security forces surrounded areas where opposition leaders and others planned to meet for a peaceful protest march against the exactions of Alpha Conde's regime.'[44]

Talking of exploitation, very few situations could match the shameless show that was enacted in the Luhwindja community in South Kivu province of DRC. The occasion was the official handing over of a secondary school that would cater for 150 students. Another project that was commissioned that day was a potable water system. What made it a show of shame is not just that the Barno Foundation, the charity arm of the Canadian gold mining company the Barno Corporation, donated the facilities. (This company's wholly owned Twangiza gold project is located close to Luhwindja.) The shame here is that this is the first high school to be built not just in the community, but also in the entire region. And the water supply would serve a population of 18,000 people in four villages. The question was, where had the government been? If it is said that the central government is far removed from Luhwindja, where had the provincial government been?

A crowd of villagers came out on 8 May 2009, probably to see government officials who had probably never visited there previously. These dignitaries included: Honourable Celestin Mbuya, DRC minister of the interior and security, the Honourable Louis Leonce Muderwa, governor of South Kivu province, Jean Claude Mashini, deputy-director to the DRC prime minister's office, Emile Baleke, president of the South Kivu Provincial Assembly, Aurelie Bitondo, former South Kivu vice-governor and adviser to the prime minister, Yav Tshibal, vice-governor of Katanga province, Colette Makila, South Kivu minister of mines, and Esperance Baharanyi, Mwamikazi (mother of the tribal chief) and member of the South Kivu Assembly for Luhwindja. Also attending were local federal and provincial members of parliament and over 15 tribal chiefs. A large delegation of Banro employees was led by company president and CEO Mike Prinsloo.[45]

What a benevolent company, Banro must be, you would be tempted to say – until you considered the total picture of their presence in the region. The company is alleged to have cut a deal with an infamous and violent militia that played a major role in the Rwandan genocide of 1994. Those who know how things work in such areas claim that Banro would have had no choice other than to relate to the militia group, known as the FDLR, or the Forces Démocratiques de Libération du Rwanda, in some way if they wanted to safely dig up billions of dollars worth of gold for themselves and their investors. Banro owns four mines in the eastern DRC province of South Kivu, a rugged landscape of jungles, volcanoes and millions of poor Congolese. Banro estimated that they could extract 10 million ounces of gold from the area and that they could make $10bn if the price of gold stayed around $950/oz. With more strict environmental legislation at home, and with funds readily coming from the Toronto Stock Exchange, Canadian mining companies head for nations where, as John Lasker put it, 'the environmental laws are weak and the politicians cheap.'[46]

The accusation against Banro is that to dig up hundreds of millions of dollars worth of gold, the company simply had to cut deals with the FDLR, a militia group that had been especially vicious in attacking the civilian population and spreading a reign of terror through murder, rape, torture, extortion, kidnappings and forced taxation. Two leaders of this group, Ignace Murwanashyaka and Straton Musoni,[47] were arrested in Germany in late 2009, while a leader of another violent group, General Laurent Nkunda, was arrested in early 2009. While Ignace Murwanshyaka and Straton Musoni appeared at the International Court of Justice in May 2011, facing charges of crimes against humanity, it was not clear whether Rwandan authorities will ever allow General Nkunda to stand in the dock. In the past, peace has been episodic in these parts because the resource wars here suck in many countries and a plethora of militia groups. With the apprehension of some of the leaders of these militias it is hoped that this cycle will be broken.

Further south, DRC is keeping alert along the borders with Angola. Maintaining good neighbourliness across borders in mineral rich regions can cost plenty of money. Early in 2009, Angola reportedly invested $13mn to increase military patrols along its border with DRC to stem the flow of illegal immigrants, some of whom came from as far away as Senegal. With a 2,000km land border, there are enough reasons to trigger squabbles between the neighbours. And Angolan troops are known to have made incursions into DRC, ostensibly to control the flow of illegal immigrants into its territory.[48]

Angola, the world's fifth biggest diamond producer, allows only companies that partner with state-run firm Endiama to explore for diamonds. Tough measures against immigrants are often taken under the cover of fighting diamond smugglers.

Far away in the US, both the House of Representatives and the Senate passed a bill to keep minerals tainted with blood from getting into the country. Recognising that minerals fuel conflicts, especially in the Congo basin, the bipartisan legislative bill[49] requires US companies to track and disclose the origins of minerals used in common electronics including cell phones and laptop computers. These include tin, tantalum or tungsten. Where the minerals originate from DRC, importers have to disclose the mine of origin.[50]

The lure of DRC gold, casserite, and coltan was so strong in the late 1990s, Uganda and Rwanda invaded with proxy militias and their own armies. Minerals readily affect peace along national boundaries. In 2000, according to John Lasker, the Rwandan military and some politicians:

> made $250mn moving coltan out of eastern DRC to Western-based mining companies and metal traders who then sold the resources to companies that manufactured parts for the likes of Sony and Motorola. Coltan, when processed becomes the powder tantalum, which is used in the making of capacitors – capacitors needed to make cell phones, video game consoles, and computers so valuable to western personal technology.[51]

Sadly, the plundering of the continent translates into direct loss of human lives, as was the case in DRC where the quest for personal electronics reportedly cost the lives of three to five million Congolese and other Africans.

While the legislative efforts went on in the US, the DRC authorities have been taking a close look at deals sealed during the 1998–2003 war and under the transitional government. A number of the agreements were signed with mining companies, and as might be expected during those difficult days, deals were easily made with less than transparent arrangements. When the government decided to review these agreements in 2007, some of the companies were unable to convince the government that they were serious about doing business. They were expected to produce, among other things, feasibility studies on their copper, gold or diamond enterprises.

As part of the mining reviews, the government cancelled the Canadian miner's Kingamyambo Musonoi Tailings (KMT) project in the country's copper and cobalt heartland of Katanga in August 2009. Tenke Fungurume (TFM) copper mine of Freeport-McMo-Ran also failed to obtain a clean bill of health in the review, and officials from the US-based firm scampered to DRC to fix things.[52]

Nigeria's Niger Delta has rightly been said to be one of the most crude-oil-polluted regions in the world today. The evidence on the ground, in the waters and in the air confirms to a casual visitor that this is the case. However, there are other toxic places that people scarcely hear about. In 2006 the mining town of Kabwe in Zambia gained infamy as one of the top ten worst polluted towns in the world, according to a report published by the Blacksmith Institute of New York. Closed in 1994, the mine and smelter are no longer in operation, but according to the institute, they:

> have left a city poisoned from debilitating concentrations of lead in the soil and water from slag heaps that were left as reminders of the smelting and mining era. In one study, the dispersal of lead, cadmium, copper and zinc in soil extended over a 20km circumference from the smelting and mining processes. The soil contamination levels of all four metals are higher than those recommended by the World Health Organisation.[53]

Kabwe, in central Zambia, first came under the spotlight in the early 1900s when the area was opened up to foreign miners through the Rhodesian Broken Hill Development Company. The mines of Kabwe yielded high-grade zinc, lead ores and vanadium. They attracted a railway line as well as a hydro power plant to make it easier to exploit the area. On account of the unregulated smelting operations that have gone on for years in the area, Kabwe has high levels of lead poisoning. The levels of lead found in children's blood in the area are up to 10 times above permissible limits.

With more than 100 years of mining, and with the transition to democratic governance, South Africa suddenly has to deal with the issues of heavy pollution from the industry. Under the apartheid regime environmental racism was a nice term and its guiding principles meant that toxic industries could be sited within racially defined lines. The same happened with toxic chemical dumps.

The South African environmental justice group, groundWork, groups like the South Durban Community Environmental Alliance and other citizen-driven organisations have worked to expose these relics of racism, demanding a clean up and possibly a clear out. A toxic tour of the South Durban area reveals very disturbing realities. In the apartment blocks close to the Shell refinery, for example, it is difficult to find a home without asthma sufferers. Indeed, kids pack their inhaler kits as others pack their lunch boxes while leaving for school. But that is just one area of a big country.

Decades of environmental contamination have affected the country's water, putting the food chain and citizens' health at risk, as discussed earlier. The closure of mines does not eliminate the problem. An admission in 2008 by a leading water researcher at the Council for Scientific and Industrial Research (CSIR) dropped like a bomb when he said: 'The truth of the matter is that as a nation we don't know how to deal with this problem because it has never happened to us before.' The researcher, Dr Anthony Turton, added that, 'This was always suppressed before because people didn't matter in the pre-1994 South Africa. All we've done so far is see the tip of the iceberg. We certainly don't have any coherent government strategies yet.'[54]

Table 4 Main waste streams in South Africa

	Council for Scientific and Industrial Research		National Framework for Sustainable Development	South African Environment Outlook	
	1992 mt/y*	1997 mt/y	mt/y	mt/y	per cent
Mining	376.0	468.2	450.0	470.0	87.7
Industrial	23.0	16.3	22.0	3.0	
Power generation	20.0	20.6	30.0	33.0	3.9
Agriculture and forestry	20.0	20.6	3.8		
General/MSW	15.0	8.2	20.0	13.5 – 15	1.5
MSW disposal			8.8		
Sewage sludge	12.0	0.3		0.1	
Total	468.0	533.6		100	
Hazardous	1.89				

Source: groundWork, 2008.[55] The figures in the table were drawn mainly from information obtained from the Department of Environmental Affairs and Tourism (DEAT).
*mt/y = million tonnes per year

One area that has received serious attention in terms of scientific studies is in south-west Johannesburg, in a valley through which runs the Wonderfonteinspruit river. This river runs from the mining town of Randfontein to Carletonville and Khutsong, and into the Mooi river. This latter river provides water for the university town of Potchefstroom. Studies show that the sediment in the Wonderfonteinspruit is contaminated with radioactive uranium and high levels of other heavy metals in wastewater discharged from local mines. These wastewaters were not treated to acceptable levels before being discharged into the water bodies. Under the apartheid regime, regulation was lax and even today enforcement of relevant laws does not appear to be enthusiastic.

Mining tops the league of waste sources in South Africa. Location of waste dumps in disadvantaged areas by the apartheid governments sowed the seed of discontent that still generates resistance today. Such dumps are found mainly in areas close to where black, coloured or Indian populations reside, according to the colour bars of the apartheid era.

A toxic tour of the Durban area is not complete without a visit to Umlazi. This is one of the so-called townships occupied by black Africans in the Durban area. A hazardous waste dump with no lining was created here in the 1980s under the KwaZulu homeland government. The dump was closed following persistent protests by residents of Umlazi, but before then and even after the closure, leakages from the dump have polluted both the Isipingo and Umlazi rivers in the area.

The sad reminder here is that there is more control and regulation on the handling of hazardous wastes in South Africa than in many other African countries. The extract, use and dump propensity of industry has rendered the African environment a toxic soup that citizens should be protected from. What we see from country to country is that following abandonments or mine closures local populations move into the hazardous pits to scrape for leftover gems and end up being exposed to grave danger. Again we see criminal negligence on the part of African leaders as well as crass opportunism by national and transnational corporations who operate below industry standards on the continent. But is this a policy failure or is there something more systemic here?

Climate chaos and false solutions

if climate change were little change
many would gather in copenhagen and exchange tips
tales to offset our elastic earth
... move movies on big screens with idiotic grins in an
age of stupidity ... stand at poisoned fires ... daring inconvenient
truths

negotiators negotiate bends open eyed
blindfolded ... listening
           with sanity canceling ear muffs

if climate change were little change
they would gather in copenhagen and exchange tips
and tales to upset our elastic earth except they
           see floods under their golden beds
and storms in their ornate tea cups
           as they gobble ice creams and artic split ice screams
turning our forests into toothpicks for their absent teeth

but if climate change were little change
           after copenhagen we will still be here

and possibly finish this poem[1]

THE WORLD CONTINUES to stand in denial of the fact that climate change is triggered mainly by society's continual reliance on fossil fuels. The larger problem is the current form of civilisation, based on Western-style consumption driven by for-profit corporations. Weaning the world off hydrocarbons cannot be postponed for ever, and it would also give us the chance to address the other major socio-economic cancers. Civilisations have come and gone not just on account of military or political conquest, but also because of a failure to adapt to challenges. Poorer nations are undeniably at the receiving end of climate change and for many, the struggle for survival is literally a swim against the tide in a turbulent, rising sea. Notwithstanding the fact that African governments and African institutions are working on coping strategies to deal with climate variabilities, a number of factors stand in the way of their ability to cope with current and unpredictable future conditions. These limiting factors include poverty, weakened institutions and fragile ecosystems.

Africa feeds herself mainly through production by her army of family farmers, especially women, representing almost 70 per cent of the population. This farming largely depends on direct rainfall, and climate change is thus a unique threat here. The continent is already persistently affected by drought. Local droughts occur every year and continental crises seem to occur at least once a decade.[2] With relentless pollution by industrial societies and the inexorable rise of global temperatures, it is very clear that the continent is being cooked on carbon fires.

The Intergovernmental Panel on Climate Change (IPCC) reckons that climate change will greatly compromise agricultural production in Africa. The panel figures that in some countries, yield from systems that are rain-fed may decline by as much as 50 per cent by 2020.[3] When climate change affects the biodiversity of the continent it has a direct impact on a people whose lives are so closely intertwined with their environment. Climate change threatens the livelihoods, food availability and health of the population.

Moreover, Africa is a continent completely surrounded by water, but suffers annually from droughts that sometimes reach crisis levels. Coastal erosion, flooding and subsidence have already affected most of its coastline and problems are bound to increase unless urgent actions are taken to build resilience.

There is no argument that with the exception of a few pockets – the Niger Delta's gas flaring by oil companies, South Africa's Eskom coal-fired electricity generators and Sasol's oil-from-coal/gas plants – Africa plays a very minor role in emissions of carbon. Those pockets aside, the continent's past economic activity did not make any significant contribution to the accumulated global stock of carbon. The main way CO_2 emissions were generated was through the extremely biased use of coal-based electricity, which was primarily directed towards mineral and cash-crop extraction, processing and transport.

As a result, whereas in other regions the key issues concern how to reduce carbon emissions, in most of Africa they concern the adaptation of production and human survival to a deteriorating ecology. This survival should logically include a higher amount of emissions, as people switch from fuelwood to electricity so as to adapt to the United Nations prediction that nine out of ten peasants will not be able to grow food by 2100. In addition, while the main adverse consequences of global warming in other regions may occur in the future, the consequences are already apparent in Africa. So adaptation is a crucial process, which our vision of climate justice should inform.

One of the factors that will play havoc with African geopolitics is the movement of human populations. Climate refugees will disdain national boundaries and seek better climes wherever they can find them. Cross-border migrations can trigger serious conflict.

For example, at one time the Burkinabe population resident in Côte d'Ivoire was up to 40 per cent of the host country's population and this contributed to the dislocation of political dynamics that eventually resulted in civil war. One of the leading figures in the political scene in Côte d'Ivoire, Alassane Ouattara, was prevented from contesting elective posts in the mid-1990s simply because he held dual citizenship – being both Ivorian and Burkinabe. The fact that he had previously been the governor of the West African Central Bank and prime minister, and had on occasions acted as president of the country, counted for nothing once the ethnic card could assure others of easier access to power.[4] He eventually became president of the country in 2011 after a standoff with the then-sitting president, Laurent Gbagbo, who refused to accept election results globally accepted as being won by Ouattara.

Already there are massive movements of people from drought-prone zones, migrations that seek to escape the famine that follows the distortion of climatic regimes. Governments wait until disaster strikes and photos of wire-thin kids and their parents flood global television networks. Rain failures on the Horn of Africa sound alarms of impending catastrophe, but these are ignored by fat cats in government houses, insulated by opulence and rings of praise singers.

There are links between population movements, due to the drought in the Sahel, and the conflict in Darfur, as pastoralists seeking water resources clash with sedentary arable farmers.[5] Climate change assures us of much more movement across borders in the future. When this hits in Africa, it often sets peoples in different camps who ought to live together and generates conflicts where there ought to be none. Climate displacements are bound to generate further conflicts in Africa, perhaps more than in any other region.

Proposals on how to tackle climate change have ranged from the ludicrous to the more realistic. One proposal is the manufacture of special anal plugs for cattle since they release a lot of methane when they fart. There were even arguments as to whether climate change was real and if so whether it was a natural and inevitable phenomenon or something to which human activities contributed. One of the key explanations offered for naturally occurring temperature rises was the output of the Danish meteorologist, Knud Lassen. He proposed that an 11-year cycle of sunspot activities on the surface of the sun matched the pattern of global temperatures. The meteorologist called for caution when he found that recent temperature surges could not be explained by the sunspot activities and solar cycles alone.[6]

In an era of disaster economics where opportunities for exploitation emerge with every crisis, the climate chaos has provided a broad play space for venture capitalists. That space was pried open at the Kyoto negotiations where the United States negotiators insisted on the inclusion of market mechanisms for tackling climate change. Indeed, they insisted that they would not endorse the protocol unless market mechanisms were given pride of place. Thus the Bill Clinton administration sent an ultimatum to the negotiators through the then vice-president, Al Gore, and successfully pressed for the Kyoto Protocol to become a set of carbon trading instruments.[7] At the end of the day, the world succumbed to US pressure and the rest is now history. The irony is that after enthroning the market as the saviour of the world's climate, the US would not endorse the Kyoto Protocol and now fights to kill it altogether.

According to George Monbiot,[8] the world as of now has no alternative to the vital need for mitigation. According to him, efforts to set the paths towards mitigation were sabotaged by the Clinton administration even before being abandoned by Bush. Climate mitigation challenges are being attended to half-heartedly by the other rich nations. In short, the global climate talks have so far totally failed. And by all accounts, the targets these talks have set bear no relationship to the science and are anyway negated by loopholes and false accounting. Moreover, where nations like the UK meet the targets set by the Kyoto Protocol they do so by outsourcing their pollution to other countries.

While countries dither about what can be done globally to avoid a 2 degree Celsius rise in global temperatures, others have learned from the Katrina floods that showed the soft underbelly of the US and that no one is immune to what is coming in the future. Thus, Germany is spending €600mn on one new sea wall for Hamburg while the Netherlands plans to spend €2.2bn on dykes before 2015. While these measures may be inadequate, they are indicators of what rich nations are capable of doing in a bid to survive the coming floods. The question is, what will happen to the poor? What is the response to the cries of the small island states such as Tuvalu whose citizens are already becoming climate refugees?

You cannot save small island states from sinking through the speculations of carbon profiteers. Actions to tackle climate change are not commodities to be auctioned or hawked in market squares or on the slippery floors of stock exchanges. The ideologues of cap-and-trade or trade before you cap are too smart by half. They lay their hands on carbon credits and then help everyone by shrewdly utilising market mechanisms to allocate emissions in the most efficient and cost-effective way. Could there be a way to code each carbon entrepreneur's carbon stock so that they could be tracked to be sure that they same stock is not being reused and resold or that those particular carbon squads are not causing particular climate catastrophes somewhere? The carbon market is the stuff of high fiction, and the best of our imaginative writers cannot begin to get near to these market big shots.

Beginning from the grandmother of market mechanisms such as the so-called Clean Development Mechanism (CDM) we are now seeing granddaughters such as the Reducing Emissions from Deforestation and Forest Degradation (REDD). A number of father and mother mechanisms fall in between, dealing with one carbon off-setting mechanism or the other. The most plausible concept behind these market mechanisms is that they provide the basis for private sector investment in efforts to tackle climate change. Less subtly put, these mechanisms provide opportunities for the private sector to control the pace and nature of projects allegedly designed to tackle climate change, but which in reality advantage vested interests while allowing primary corporate polluters to evade accountability. This explains why these schemes for carbon offsets really upset some people and communities.

There is no disputing that the upsurge in global temperatures is due largely to the amount of carbon released into the atmosphere. Carbon is a basic building block in every living thing, plant or animal. Our soils are loaded with carbon and so are our air and oceans. We take in oxygen and exhale carbon dioxide. Plants do the reverse and we coexist happily supplying each other's carbon dioxide or oxygen needs. The problem is that over the last two centuries humans have dramatically increased the amount of carbon dioxide and other greenhouse gases released into the atmosphere. It is estimated that about 26 billion tonnes equivalent of CO_2 is released this way yearly. One way to visualise how CO_2 plays a vital role in the climate equation is to see the gas as enveloping the entire earth, occupying a belt around it, so to speak, in the lower atmospheric regions. There are several other gases there, but the essential difference is that this gas, along with other greenhouse gases, allows the energy from the sun to reach the earth but slows down the escape back into space of the energy reflected from the earth's surface. This build-up of energy is the greenhouse effect.

The question that confronts humanity is how to tackle this carbon in the atmosphere. What should people do? Do we stop releasing carbon into the atmosphere? What would that entail?

Are there acceptable levels that we can keep to so as not to reach the point of no return with a runaway climate change? These questions have engaged the imaginations of many and a variety of solutions have been offered.

One interesting proposed solution is carbon sequestration. In this scenario you could continue with business as usual, releasing as much carbon as you please, only ensuring that the carbon is captured and stored or sequestered. Technologists have been at work on carbon sequestration technologies, with some already undergoing tests but with none likely to be ready for practical use until about 2020.[9] Meanwhile, carbon offsetting is gaining ground in some quarters with the proposition that one could pollute in one part of the world, and then have the pollution offset in another part of the world. For example, a company in Europe could keep on stoking the atmosphere with carbon, but plant a tree plantation somewhere in Africa and since the trees absorb carbon, the company can feel satisfied that they are carbon neutral. It is a convenient fictional scenario. The company cleans up its conscience; those setting up the plantation get paid. Communities whose lands are taken up to set up the plantations may get jobs as plantation hands and perhaps receive some form of compensation for lost farmlands. What the communities can be sure of is that their lands will slip out of their grasp.

The CDM can sometimes be applied to ridiculous extremes. Take, for example, the marketing of gas flare stopping projects as CDM projects. Through such projects, oil corporations and the Nigerian government hope to claim carbon credits for helping 'fight' climate change. The reality is that gas flaring has been an illegal activity in Nigeria since 1984 when the Nigeria law on gas re-injection came into effect. Any reduction or stoppage of flaring is simply a reduction or halting of a criminal activity and brings about no additionality such as the CDM process requires. Any compensation for such an activity flies in the face of reason. Gas flares are the most cynical manifestations of corporate insolence in the face of climate change and environmental health. The flares release greenhouse gases such as carbon dioxide, methane and nitrous and sulphur oxides. Apart from these, the flares release other harmful substances that greatly affect human health.

There is no certainty about how much carbon a particular tree absorbs. In any case, when the tree eventually dies, where will the carbon go? Offsets allow industrialised nations to take no step to halt their polluting actions, including consumption reductions, at home while undertaking hypothetical cuts elsewhere in the developing world. Although developing regions like Africa may receive some economic incentives through the offset markets, there is no indication that the amounts transferred are adequate, fair or enough to mitigate the relentless pollution going on in the north.

According to Carbon Trade Watch, what offset companies do is nothing more than selling peace of mind to consumers, thus breeding complacency. They also posit that the most polluting corporations and politicians use offsets as a cheap means of greenwash, 'as distraction from their inherently unsustainable practices and refusal to take more serious action on climate change'. In addition, creative accounting coupled with 'dubious scientific methodologies' are used by some offset practitioners to inflate profit. Furthermore, our knowledge of the carbon cycle is limited and it is not possible to say whether plantations have a positive benefit in terms of mitigating climate change, 'let alone exactly quantifying the supposed benefit into a sellable commodity'. Finally, it is impossible to determine a hypothetical baseline from which to measure what would have happened if the carbon offset project was never embarked on.[10]

We also hear of cap-and-trade systems. These have been characterised as systems by which governments issue licences or permits to allow polluting industries to pollute. As critical researchers Tamra Gilbertson and Oscar Reyes remark, 'Instead of cleaning up its act, one polluter can then trade these permits with another who might make "equivalent" changes more cheaply … In practice, the scheme has failed to incentivise emissions reductions.'[11]

Before we turn to the justice aspect of climate change, we cannot avoid bringing up the question as to why humans are avoiding taking real actions, embracing real solutions to a real problem, but rather endorsing a series of false solutions. It is like playing the clown in the face of tragedy or making jest in the house of mourning. If increased stocks of carbon in the atmosphere cause the challenge, the real action must be to drastically reduce the release of more carbon into the atmosphere. If the release of carbon is mostly through the use of fossil fuels, then the solution must lie in a radical shift from this driving force. If carbon sequestration is the way to go, then the best way to ensure this is to leave the crude oil or bitumen in the soil and the coal in the hole, and to flee the gas fracking business. It is so simple that it is almost stupid not to see it.

Blind climate negotiators

Climate negotiations, from Durban in late 2011 onwards, will increasingly confront the issue of climate justice. The atmosphere is a common space, a global commons. Industrialised nations pumped a disproportionate amount of emissions into the atmosphere and they have cornered a disproportionate amount of global resources, largely by exploiting nations that are on the other side of the coin. Climate impacts are already being felt in a severe way in Africa as well as in other regions of the global South. Centuries of exploitation have weakened the resilience of these regions and in tackling climate change these historical facts must be addressed. One way of addressing this is by the payment of climate debt to make the needed financial and technological resources available to these vulnerable regions.

The Conference of Parties at Copenhagen and the following one at Cancun did not generate outcomes consistent with scientific warnings that the world faces a severe climate crisis. Copenhagen ended with an accord spearheaded by President Barack Obama of the United States with the backing of the BASIC countries (Brazil, South Africa, India and China) concocted in a 'Green Room' dreamed up by Denmark's conservative ruling party. In that room, Patrick Bond recalled, were 26 countries 'cherry-picked to represent the world. When even that small group deadlocked, allegedly due to Chinese intransigence and the overall weak parameters set by the US, the five leaders (Obama, Lula da Silva, Jacob Zuma, Manmohan Singh, and Wen Jiabao) attempted a face-saving last gasp at planetary hygiene.'[12]

The demand of climate justice is that those who created the climate problem must be the ones to mitigate it, and in the process must transform their economies and societies.[13] There are two ways to go about making this happen. First, rich nations must reduce rapacious consumption patterns and address the climate crisis with real solutions and not ones that have been seen to be false. Second, the rich nations have to support the poor nations who are being forced to adapt to a situation they did not create. One practical way of making that happen is through support for sustainable, green development paths.

Among governments, the Bolivians have made the clearest call for climate justice while India and China have used related arguments to defend their growth paths. At a time when the world has been calling for a curtailment of polluting industrial establishments, China has been building new coal-fired power plants at a prodigious rate.[14] It is interesting to note that while China is massively expanding its coal-powered plants, it is also quickly assuming leadership in the utilisation of wind power. The discourse on how much both China and India must do in tackling global warming must not overlook the fact that vast numbers of people in both India and China still require electricity supply and that meeting that gap requires huge financial outlays.

Following the catastrophic outcome of the United Nations climate negotiations held in Copenhagen in December 2009, President Evo Morales of Bolivia announced that the world would meet in Bolivia for a thorough and inclusive discussion on this vital issue.

The summit, held in Cochabamba in April 2010, attracted 35,000 participants from 140 countries. The summit stood in sharp contrast to the Copenhagen event in many ways. First, this was an assembly of governments and peoples. In Copenhagen no effort was spared in keeping civil society out of the conference: the conference was marked by lockouts of civil society, detentions of climate activists and outright brutality towards non-violent protesters on the streets. In Cochabamba the police were offering assistance and were also participants. Whereas Copenhagen showed a disdain for the voices of the people, Cochabamba was about raising the voices of the people. The only similarity between the events is that they were both held in cities whose names start with letter 'C' followed by nine letters.

The key outcome of the Cochabamba conference was the People's Agreement. This agreement demanded that countries cut their emissions by at least 50 per cent at source in the second commitment period of the Kyoto Protocol (2013–17), without recourse to offsets and other carbon trading schemes. In terms of finance, the People's Agreement demands that developed countries commit 6 per cent of their GDP to finance adaptation and mitigation needs. The financial suggestions of the Copenhagen Accord are a drop in the ocean compared to what is needed to secure vulnerable peoples and nations. The peoples of the world also affirmed that there is a climate debt that must be recognised and paid. The payment is not all about finance but principally about decolonising the atmospheric space and redistributing the meagre space left. Developed countries already occupy 80 per cent of the space.

The climate debt is also about taking actions needed to restore the natural cycles of Mother Earth and one clear way of achieving this will be through the proclamation of a Universal Declaration on the Rights of Mother Earth, with clear obligations for humans. Bolivia is in the forefront of promoting the adoption of this declaration at the United Nations. The People's Agreement recognises that the causes of climate change are systemic and that systemic changes are needed to tackle them. On this note, the model of civilisation that is hinged on uncontrolled development can only compound the crisis. The world needs to move towards living well and not continue on the path of domination of others and of conspicuous and wasteful consumption.

An area glossed over in the United Nations Framework Convention on Climate Change (UNFCCC) negotiations is the role of industrial agriculture in climate change. The People's Conference debated this key sector and reached the agreement that the way to a sustainable future is through the enthronement of food sovereignty based on agro-ecological agricultural systems. The issue of access to water being a human right was also affirmed by the people and later on in the year by the United Nations.

In all, the People's Agreement recognises that real strategies to tackle climate change must be based on the principles of equity and justice in dealing with the structural causes. Without climate justice it will also clearly be impossible to achieve the much talked about Millennium Development Goals (MDGs).

Cochabamba resonated with calls for urgently securing the rights of Mother Earth as a means of reconfiguring our

Table 5 Per capita CO_2 emissions for selected countries 2005

Country	Per capita CO_2 emissions in tons
United States	19.6
Australia	18.4
Japan	9.5
China	3.9
India	1.1

relationship with the earth and with each other – in a way that respects the past, today and the future. All these will be a pipe dream unless peoples' sovereignty is supported, restored or built across the world. Cochabamba was a turning point in the march to transform our world from the path of conflict, competition, exploitation and domination to a path of solidarity and dignity. It held a ray of hope for Africa.

With approximately 10 per cent of the world's population, the United States and the countries of the European Union contribute more than 50 per cent of the carbon emissions in the atmosphere. It is useful at this point to compare the per capita emissions of selected countries in the world (see Table 5).

With regard to consumption levels, the argument has been raised that the enormous growth of the middle class in both China and India has led to a rise in the demand for beef, which in turn means dependence on intensive cattle production systems that results in more deforestation as well as the use of more polluting agro-chemicals. This argument may be material for stand-up comics, but the reality is that the meat industrial complex has enormously contributed to the release of greenhouse gases through deforestation and land conversion to ranches as well to soy plantations in South America, for example.

In fact, while the US has a mere 5 per cent of the world's population, these over-consumers emit nearly 25 per cent of the world's greenhouse gases from the burning of oil, gas, and coal – for driving cars, producing electricity, and running industries. This huge carbon burden may have been a major reason why the country would not readily want to accede to emission caps that would help keep the earth's temperature from rising to or above 1.5°C over pre-industrial levels. Already the earth has warmed by almost 0.8°C since the Industrial Revolution. The IPCC in its fourth report estimated that at this rate, temperatures will rise by 2–2.4°C by 2050. To avoid this, greenhouse gas emissions must be cut by 50–85 per cent relative to 2000 levels by 2050, and if nothing is done to check the rise in temperature, up to 30 per cent of plant and animal species will be under threat of extinction.

This point bears repetition. The market ideology bent of the Kyoto Protocol permits countries unwilling to achieve the target set by the protocol to continue with their emissions binge and notionally offset these sins by some other mechanisms:

- Buy emissions rights from countries who do not exhaust their emissions quota.
- Invest in forestry and soil conservation projects elsewhere, with the understanding that such projects would lead to absorption of carbon to equally compensate for their continued emissions.
- Invest abroad in projects that would save on greenhouse gases. Such projects are known as Clean Development Mechanism (CDM) projects and are often carried out in countries that do not have obligatory limits. They can also invest in Joint Implementation (JI) projects in other industrialised countries.

The major thrust of carbon trading and carbon offset strategies is to transfer the responsibilities for the impacts of climate change to the South while the polluters reap profits from the new business built upon disasters. Advocates of carbon trading make the argument that this will allow economies to adjust more rapidly and efficiently, with fewer disruptions. In other words, actions that ought to call for penalties are now overlooked because of some queer financial mechanisms that leave the environment at the mercy of powerful polluters. Carbon trade and other false solutions such as genetically modified (GM) organisms, carbon sinks, ocean fertilisation, carbon storage and agrofuels, are all formulas that leave aside the oil industry, the number one sector responsible for global warming.[15]

As already noted, the climate crisis is being seen as an opportunity to force unpalatable solutions on unsuspecting populations.

The orchestration of the 2007–08 phase of the food crisis offered proponents of genetically modified crops a golden opportunity to say this was the time when hungry Africans could not say no and, indeed, should not say no to any kind of food generously placed on their shaky dining tables. The unfortunate thing about these things called GM crops, as with other proposals for the continent, is that many African governments tend to see them, first of all, as opportunities to receive grants. The issue of whether the technology is suitable to the African context, including our health and eating habits, is secondary.

'Africa needs technology transfer.' 'Africa can domesticate the technology.' 'We will not be tied to the apron strings of Monsanto.' 'A billionaire like Bill Gates, who is sponsoring the Alliance for Green Revolution in Africa (AGRA), cannot be wrong.' 'GM crops will help African farmers increase yields, grasp the value added chain and penetrate global markets.' The genetic engineering train is presented as a silver bullet that solves all the imaginable agricultural and food problems of the continent. And African leaders ululate in wonder.

Africa has become a major battleground for GM crops, and efforts to ensure the penetration of the continent by hook or by crook have been thick and persistent. With a picture of unyielding hunger, malnourished people and an inability to plough with anything other than a hoe, Africa is presented as a lost cause that must be helped by a loving, caring world. Climate change provides a wonderful cover to push this agenda. The GM industry claims they can produce varieties that do not need water to grow and indeed will produce others than do not require agrochemicals. The promises keep rolling off the tongues of their salesmen. The message is that without GM crops and with climate change, Africans will not only be left with empty bowls, but the continent will become a dust bowl.

True, the effects of climate change are clearly manifest in the agricultural sector. But, trust the purveyors of genetically engineered crops. They want to gain from all ends of the pipe. As we often hear, agriculture contributes a large chunk of greenhouse gases into the atmosphere and thus impacts on the climate. What is often not heard is that the culprit is industrial, chemical-dependent agriculture and not the environmentally sound agro-ecological practices. Nevertheless, even though smallholder farmers are not climate criminals, they are severely impacted by it. For the Niger Delta, continued degradation in the form of spills and gas flares render the area extremely vulnerable to the impacts of climate change with a projected loss of 50 per cent of the ability to produce cereals by the year 2020 and an 80 per cent loss by 2050.[16] This is worse than any armed conflict.

In grasping at carbon credits and offsets, some entrepreneurs suggest that, for the agricultural sector to benefit from the carbon market, systems that reduce the need for tillage should qualify. If this is taken on board, genetically modified crops such as Monsanto's Roundup-ready varieties could be used to claim carbon credits by arguing that they reduce the need for tillage and thus reduce emissions. This would be a perverse incentive for growing GM crops.[17]

Seeing REDD

The REDD debate, like much else in the climate talks, is laced with political-ideological undertones, commercial interests and a lack of attention to the realities and challenges of communities that stand in the line of fire from the volleys of these talks. Making forests a commodity through reducing emissions from deforestation and forest degradation (REDD) is another ingenious move to avoid putting in place real solutions to the climate problem. REDD does not tackle the root causes of deforestation. It simply provides a cover for unabated pollution in the North while buying up forests, and forest carbon credits, in Africa and other forest-rich regions.

About 20 per cent of carbon emissions annually come from deforestation. Of the 1.6 billion people who rely on forests 60 million are entirely dependent on forests for medicines, foods, building materials and other means of livelihood. REDD will probably make these forest-dependent people see red, what with the new management systems that will emerge with the privatisation of forests and the exclusion of forest communities from the deals. We can expect the conversion of forests into monoculture plantations with huge social and economic impacts on local communities.

In the REDD negotiations so far, cracks have appeared in the positions of developing countries. While most are thinking of the financial gains that may arise, others are struggling to ensure that governments retain control – to the exclusion of the people who are the ones living and depending on the forests – over decisions on what uses can be made of forests. What this means is that while Africa is still burdened with centuries of exploitation, governments across the continent have continued, in large part, to replicate the same destruction through the vehicle of 'neocolonialism' posing as market democracy.

Some of these issues emerged very clearly in the negotiations that took place in Bangkok in October 2009. A report[18] from the Third World Network on the debates of 8 October is instructive as regards African nations such as Gabon. The contentious debate revolved around the conversion of natural forests to other uses, and the rights of indigenous peoples:

> The Ad-hoc Working Group on Long-term Cooperative Action has an informal sub-group working on paragraph 1(b)(iii) of the Bali Action Plan. The focus is on policy approaches and positive incentives on issues relating to Reducing Emissions from Deforestation and forest Degradation (REDD) in developing countries; and the role of conservation, sustainable management of forests and enhancement of forest carbon stocks in developing countries. Currently, only reforestation and afforestation activities are included as part of the Clean Development Mechanism (CDM) under the Kyoto Protocol. Avoided deforestation does not qualify under the CDM as defined now.
>
> The second week in the Bangkok talks that end today concentrated on developing a non-paper regarding the safeguard principles for an outcome on REDD-plus ... After a number of contact group meetings and informal meetings... the consolidation of text seemed to be unsuccessful in incorporating all views.
>
> For **Gabon** a principle based approach to respecting indigenous people's rights is one that accords with the law of the country. It explained that a concept of rights that provides specific rights to people is risky. In the same way we talk about the indigenous people of forests, what would we do if people of the oceans wanted specific rights to the oceans, and other peoples wanted specific rights to other areas? This would be a problem.

Democratic Republic of Congo also on behalf of **Cameroon, Republic of Congo** and **Equatorial Guinea** said that their countries were still heavily forested and that there is still rampant poverty. Thus they still need to have the sustainable exploitation of the forest to meet their people's needs. They explained that 35 per cent of their land mass had been protected, but outside of these areas they need to use forest for economic development, unless they get adequate compensation to protect them. They argued that reference to land conversion should not be returned to the text by countries that have already widely logged.

Liberia argued that while they were fully committed to REDD and support in principle the safeguards in the text, they stated that 'two wrongs don't make a right' and thus they have national laws that they must oblige by, in respect to logging rights for companies. So it cannot be a Party to a convention that is not legal in its laws. It supports not converting land; however we need to find a way to right the wrong while respecting national laws. It needs to protect future business confidence. So countries need compensation and financing related to governance.

All said, REDD is simply another mechanism to transfer the responsibilities for tackling climate impacts to the South, creating new threats for the people, including conversion of indigenous territories into plantations, land grabbing and displacement of populations. These mechanisms provide a cover for forests to be given to private businesses and equally aid the privatisation of protected areas and natural forests, the occupation of peasant and agricultural lands, and the deprivation of the local communities of their rights and livelihoods. All these mean a subsidy to the polluter/business and a stimulus for energy-guzzling countries in the North to maintain their production and consumption models.

We can say that REDD is at least honest in an aspect of its intentions. It is aimed at reducing deforestation and does not seek to stop it. What forest people want, and what the world needs, is an end to deforestation. Likewise, those of us in the Niger Delta and many other sites of oil exploitation need an equally principled position to emerge, for our own sake and for that of the planet: leaving fossil fuels underground.

7

Leaving the Niger Delta's oil in the soil

This drilling
This killing
This stealing
This maiming
This raping
This spilling
This desecration of Papa's land now we ask as the singer did:
Who owns Papa's land?
Justice now!
No reconciliation ... without justice[1]

THE WORLD KNOWS that the climate challenge is caused primarily by the release of carbon into the atmosphere through the burning of fossil fuels. The question is: why is the world unwilling to tackle this problem at the roots? Why are policymakers pretending they are not aware that the current dirty energy models will keep compounding the problems and that the problems cannot be offset no matter how much money is invested in the market over hot air and no matter how innovative we become in doing that?

The simple answer to our climate crisis, one begging to be accepted, is that we must simply leave the oil in the soil, the coal in the hole and the tar sands in the land. We do not require expensive carbon capture and storage technologies to make this happen. It is just common sense. Simple.

As sensible as leaving the crude in the soil is, the world's addiction to crude is pushing the oil majors and minors to move even deeper into fragile ecosystems, into deeper waters and into dirtier forms of fossil energy sources. A new oxymoron, clean coal, is accepted without batting an eyelid. Fracking shale gas is alleged to be cleaner than coal or oil, though it typically is not, given how much mess is required to get the gas.

In the search for new oilfields and for the securing of present ones, Africa stands ready to be poked full of holes. Although Africa boasts no more than 10 per cent of proven world oil reserves, the continent is the desire of oil guzzling nations of the North, notably the United States and increasingly China. NATO's attack on Libya is an example of the extreme measures that will be taken. And although Africa stands little chance of replacing the Middle East as the oil centre of the world, more rigs are clawing into the continent and more oil tankers and floating production and storage facilities are rapidly springing up.

For some time, Nigeria held 12th position in the league of the highest producers of crude oil in the world. The country's production quota has generally been put at 2.2 million barrels per day. In 2009 her oil production quota as agreed by OPEC was 1.6 million barrels per day. We note here that even that reduced target could not be met for several months. The fact that Nigeria did not meet the production quota, however, did not necessarily mean that she had not been producing up to or above the quota on a daily basis over the period. It is common knowledge that oil thefts grab anything from 250,000 barrels to an equal quantity to that formally exported every day, going by official estimates and estimates of those who should know. This means that probably up to 2 million barrels are being stolen every day. Indeed, a speaker of the Nigerian House of Representatives estimated that the amount of oil stolen on a daily basis in the Niger Delta was equal to the amount of crude oil being legally exported daily from the country.[2]

Simple reasoning would suggest that should the Nigerian government wish to increase oil production, the first step must be to halt oil thefts. In fact if oil theft were stopped there would be no need to raise production levels. If that were done, the government would immediately add an extra one or two million barrels a day to national production levels. To achieve that feat, the Nigerian government would require massive political will to halt the prodigious stealing in the oilfields. This will is in deficit because it is suspected that those engaged in the stealing are not small fry, but rather men who occupy high positions in society, are well connected and have their backs covered. Many imagine that oil thefts are executed with buckets and bottles. Such thinking is fed by stories of pipeline fires and abandoned buckets and jerry cans. The poor get trapped in such fires and the authorities use that as an excuse to criminalise local communities rather than accept their failure to maintain and protect the pipelines and petroleum product depots.

A very interesting example of the high-calibre thievery going on in Nigeria is highlighted by the case of the MT *African Pride*, a Nigerian oil tanker vessel that was caught near Shell's Forcados export terminal with 11,000 authorised barrels of crude oil in late 2003. It had on board 13 Russian crew members, who were arrested by officers of the Nigerian Navy. In August 2004 it was suddenly discovered that the MT *African Pride* had slipped out unnoticed from custody. When it was tracked down it was found that its cargo of 11,000 barrels of crude oil had been transferred to another vessel while its hull was loaded with seawater. How could a ship slip out of naval custody without being noticed? That is something to think about.

With so many unknown factors in the Nigerian oil sector, it is interesting that the country bothers to forecast how much oil they have in reserve and for how long it might last. It is true that governments do have an enormous amount of information that citizens do not have, but if no one knows how much crude oil is being extracted, such estimates of proven and unproven reserves are fancy talk. The amnesty offered belligerent gangs in the oilfields could ensure the stepping up of oil production, but will it halt the thieving? Would it stop the reckless and massively destructive extraction that has been the norm over the decades?

In recent years, some peace did return, as was hoped. When the amnesty was announced it was hoped that it would be through a holistic, open and inclusive dialogue with a broad spectrum of citizens in the region and the nation at large. The huge budget for the amnesty appeared at a point to simply be a way to feed the troops and dig more trenches in the region. Except for some infrastructure projects, the funds would certainly not address the root challenges of the Niger Delta environment.

On 6 September 2011 a massive protest by persons describing themselves as ex-militants blocked the arterial highway that links western Nigeria to the east through the Niger Delta. The response of the special adviser to the president of Nigeria on the Niger Delta and the 'Chief Executive Officer of the Presidential Amnesty Programme' reminded the belligerent youths in a statement that :

> on June 25, 2009, the Federal Government of Nigeria proclaimed amnesty for specific persons in the Niger Delta with a view to resolving the protracted insecurity in the zone. The terms of the amnesty included the willingness and readiness of agitators to surrender their arms, unconditionally renounce militancy and sign an undertaking to this effect. In return, the government pledged its commitment to institute programmes to assist the disarmament, demobilisation, rehabilitation and reintegration of the former agitators.[3]

It appears that more and more youths have sought to be registered as ex-militants so as to benefit from the financial perks as well as overseas training programmes that some ex-militants have enjoyed. Some have been sent to countries like South Africa, Malaysia and elsewhere to acquire skills in oil pipeline welding, among others. Pipeline welding is a poetic skill for persons suspected of having been engaged in sabotaging pipelines. We note the emphasis in the statement that the amnesty was offered to 'specific persons'. The reality is that with an army of unemployed youths in the region and with burgeoning violent actions elsewhere in the country, more people may seek to be included in the category of these 'specific persons.' Will the cycle continue?

Nigeria and other African nations have unfortunately been caught in the trap of global financial and economic crises. An expected fallout of the crises is the shift to more rapacious extraction of natural resources in order to meet expected shortfalls in income levels. Countries will dig for more gold, diamonds, copper and oil. Besides simply printing currency notes, the push for massive natural resource extraction always beckons leaders who have no time to sit back and plan for the liberation of their peoples. This path portends more conflict and environmental degradation – extract more timber, catch more fish, mine more. The drive goes on.

Africans need soil, not oil. The environment is the cradle in which Africans are nurtured. Crude oil extraction has effectively uprooted the people from the soil. It has polluted their waters and poisoned their air.

While tensions persist in the oil regions of the continent and with looming peak oil – the view that maximum production from global oil reserves has already occurred, and can now only be followed by a terminal decline – perceptive nations can only look forward with trepidation. In the midst of the tensions, oil corporations operating on the continent continue to garner obscene profits. This happens because the corporations are not paying for the environmental costs of their operations and because ecological debts go unattended to. Local communities continue to shoulder the externalised costs while the corporations laugh all the way to the banks, secured by their opaque joint venture and production sharing agreements.

The trend of profits made by oil companies over the past couple of years is indeed very telling. These companies reap profits in the face of whatever woes the world is confronted with. In 2007 Shell's net profit rose to $11.56bn from $8.67bn a year earlier.[4] According to reports, Exxon, the world's largest privately held oil company, reported a 14 per cent rise in profit to a record $11.68bn, which was adjudged to be the largest ever for a US corporation. In the first quarter of 2008, Exxon made nearly $90,000 profit a minute![5]

The convulsions gripping the global system have directly affected the economic outlook of African nations and especially Nigeria. At a point the major challenge of the Nigerian state was related to the collapse of crude oil revenue from an unprecedented height of about $150 a barrel to below $40 a barrel. Gains were made later on in 2009 when the price climbed to over $50 a barrel. And the drums were pulled out for celebration. The crash revealed that behind the cheap piles of petrodollars lies systemic inequality underpinning the 'free market' with its inbuilt disadvantages. And the so-called market forces are not as free or invisible as international financial institutions would want the world to believe.

Some Nigerians are equally worried that even the cheap oil that the nation depends on may soon be set aside due to the real possibility that the world will move on to new alternative energy sources. If that happens and crude oil attracts less attention, what will be the consequence for the Nigerian economy? While these are legitimate concerns, they also present a great opportunity to transform the environment and by extension the economy.

Cheap petrodollars drove Nigeria into believing that money was not a problem; rather, it was how to spend it, as said a former national leader. The leaders drove the country into debt and debased her sense of nationhood. Cheap petrodollars turned Nigerian politics into a struggle for the control of the national purse and led to a regime where there was a massive move of public funds and properties into private control. That has been the visible meaning of privatisation in the nation. Cheap petrodollars invited the jackboots in and rocked and overturned every sense of common good and collective owner-ship in the nation.

The drive to maintain the flow of foreign exchange into national coffers made it impossible for the government to see that a safe environment is a basic requirement for citizens to be productive. African governments overlook the fact that in largely subsistence economic systems where the vast proportion of the citizens thrive outside of the formal economy, the first thing that must be secured for national health and productivity is an envi-ronment that supports the people's efforts in the areas of family farming and livelihoods. The grave inability to grasp this truth has allowed oil companies (national and transnational) to operate with impunity in the oilfields and to pollute, destroy and dislo-cate the very basis of survival of the people in Africa.

There is a clear proposal on how to turn the crises into a real opportunity to break from an ignoble system and move on to a sustainable path. As they say, dropping shackles always requires sacrifice, especially if it involves jettisoning fear, divisions and firmly held prejudices.

The issue of gas flaring is a burning one that must be addressed as a matter of urgency. An estimated 168 billion cubic metres of natural gas is flared yearly worldwide and 13 per cent of this is flared in Nigeria (at about 23 billion cubic metres per year). Flares lick the skies of Equatorial Guinea, Democratic Republic of Congo (DRC), Angola and other countries with oilfields. The many health impacts of gas flaring are well documented and include: leukaemia, bronchitis, asthma, cancers and other diseases.

In economic terms, Nigeria, for example, sends over $2.5bn worth of gas up in smoke annually, going by 2005 estimates. If we assume that this rate held good for say 10 years, we are talking of $25bn wasted and if we extend it to the past 20–50 years that figure multiplies. For each additional year that the government refuses to act the amount wasted continues to rise, as does the log of the dead due to the poisonous nature of the gases.

We are worried that at a time when the world is seeking ways to combat global warming, oil companies are busy cooking the skies through gas flaring. From pronouncements on climate change emanating from Nigerian government agencies, it is obvious that the government cannot plead ignorance of the massive contributions of gas flaring to global warming. There can be no excuse for this unhealthy and uneconomic act.

A quote from a 1963 confidential communication from the British trade commissioner to the UK Foreign Office shows that oil corporations have been engaged in this action for at least half a century now. The 50-year-old script of pacification requires urgent critical political, environmental and socio-economic examination and replacement:[6]

> Shell/BP's need to continue, probably indefinitely, to flare off a very large proportion of the associated gas they produce will no doubt give rise to a certain amount of difficulty with Nigerian politicians, who will probably be among the last people in the world to realise that it is sometimes desirable not to exploit a country's natural resources and who, being unable to avoid seeing the many gas flares around the oilfields, will tend to accuse Shell/BP of conspicuous waste of Nigeria's 'wealth'. It will be interesting to see the extent to which the oil companies feel it necessary to meet these criticisms by spending money on uneconomic methods of using gas.

It was not until the 1979 Associated Gas Reinjection Act that routine gas flaring was finally outlawed in Nigeria. Section 3 of the act set 1984 as the deadline after which companies could only flare gas if they have field-specific, lawfully issued, ministerial certificates. There are several flare sites still emitting a toxic mix of chemicals into the atmosphere in the Niger Delta. Through this obnoxious act the country has lost about $72bn in revenues for the period 1970–2006 or about $2.5bn annually.[7]

The Nigerian Senate passed a Gas Flares Prohibition Bill in 2009, setting the penalty for gas flaring at the market price of gas being flared. While this was a good move, what was needed was an order for the immediate stoppage of gas flaring, even if it meant shutting down the offending oil wells. However, the Senate bill did not become law as the House of Representatives stalled and made no progress on it before their tenure expired in May 2011.

Interestingly, in 2009 the government of Chad banned the use of firewood and charcoal as cooking fuel to combat desertification and deforestation. Citizens were expected to use gas cookers rather than firewood hearths. While that may be a great theoretical solution to climate challenge, the question is: did the law take cognisance of the economic realities of the citizens. Were the alternatives accessible and affordable? What investment was the government making to improve the lot of the people? Except for a refinery which the Chinese are to build close to Ndjamena, all the petroleum currently being extracted from Chad is piped out of the country through ExxonMobil's Chad–Cameroon pipeline, financed by the World Bank. All measures must be taken to combat the health impact of using inappropriate fuels. And the first is not to quench the fires beneath our cooking pots; it is to review the use of available national wealth and halt activities that char the livelihoods of the people.

The stoppage of gas flaring will mark a major step towards detoxifying the Niger Delta and other oilfields in Africa. The other steps that need urgent work are twofold. First is the immediate auditing of all oil spills, drilling mud and cuttings discharges, produce-water handling and other related polluting incidents in the entire Niger Delta. Second is the immediate commencement of a thorough clean-up of the environment to international standards such as those set by WHO for safe drinking water and air quality.

These steps would make it possible for the people to farm and fish with reasonable hope of achieving living incomes from such activities. Life expectancy would also increase beyond the current 41 years, as the environment would once more become people friendly.

From a report of the assessment of the environment of Ogoniland prepared by the United Nations Environment Programme (4 August 2011) cleaning the polluted waterways will take at least 30 years. It is akin to issuing a death certificate for the territory.

There is no future in crude oil as the major revenue earner or even energy source. The future of crude oil is already history. As a continent that is most vulnerable to the impacts of climate change, Africa should lead the way by not making any new oil block concessions. Meanwhile, existing fields could continue to be exploited, while alternatives are vigorously sought. Halting the giving out of new oil blocks would not mean a major loss in revenue. African governments need to review current oil contracts and see if they are actually deriving any decent income for all the impacts being suffered by the communities in the line of fire. A test question for African governments should be whether they believe that the revenue they derive from crude oil sales could actually clean up and restore the environments affected by oil extraction. If the answer is no, or even a maybe, the illogicality of those who proceed with the extraction should be seen to their shame.

The oil under the ground is still Africa's oil. Africa must *not* exploit every resource simply because it is there. This is simple wisdom.

Peak oil is a reality of geology. Quite soon a peak in production will be reached, with the world having used more than half of all currently proven reserves.[8] It is already estimated that Nigeria reached her own peak oil level a couple of years ago. The plan to increase production levels of about 2 million barrels of crude oil per day to 4 million barrels per day in 2010 or 5.2 million barrels a day by the year 2030 can be done without sinking any new wells.

Economic considerations or how this can be done

Environmental Rights Action/Friends of the Earth Nigeria (ERA) has made a proposal on how to leave Nigerian oil in the soil without causing an upheaval in the national treasury. The organisation suggests that this proposal can be replicated for any African country and can be adjusted to suit local realities.

It begins by assuming that Nigeria would have probably been in a position to increase her crude oil production from 2015 by, say, 2 million barrels a day from new oil blocks, which it demands

125

should not be given out to the bidders. If current rapacious oil thefts were stopped, the nation would have those 2 million barrels right away. Assuming that only 1 million barrels a day would be recovered from the thieves, then the nation would be seeking 1 extra million barrels a day by 2015.

By keeping new oil in the soil, Nigeria would have kept the equivalent tonnes of greenhouse gases out of the atmosphere. This would be a direct measure of curbing global warming through an infallible technology of carbon sequestration. This is a foolproof step that requires no technology transfer and does not require any international treaty or partnership.

ERA recognises that were Nigeria to trade that amount of carbon using any of the available market mechanisms for tackling climate change, such as the so-called Clean Development Mechanism, the country would earn some good income from keeping the oil under the ground. But ERA does not support the use of market mechanisms for such purposes. It rather suggests the halting of crude oil thefts and the massive capital flight from Nigeria would boost the economy and offset whatever might be seen as a 'loss' of projected revenue from crude. Climate debt paid by the industrialised world to the areas most adversely affected by climate change would be the other crucial strategy to ensure areas like the Niger Delta are cleaned up at the expense of the beneficiaries.

But even without those payments there is a strong logic for leaving the oil in the soil. ERA adopts a conservative pricing assumption, namely that if crude oil prices stabilise at $60 a barrel over the next several years an additional 1 million barrels a day would bring in daily revenue of $60mn or an annual income of $21.9bn. Now, assuming Nigeria's population stands at 140 million, it would mean that each citizen would need to buy $156.4 worth of crude oil per year.

If we factor in production costs (including staff salaries, payment of the military, etc) and company profits, we can safely say that the amount that would get to each citizen would be much less than $156.4 per year.

The organisation proposes that rather than exploit new oilfields with the attendant pollution, human rights abuses and malformed political system, the nation should keep the oil under the ground and require that every Nigerian pays a mandatory maximum $156 a year into a crude oil solidarity fund or tax. This would bring additional revenues to whatever the country makes from current oilfields, including the corked ones.

ERA members naturally recognise that not every Nigerian can afford to pay $156 a year into the national coffers, but expect about 100 million Nigerians to enthusiastically make this payment if the benefits are carefully made public. Those who can pay multiples of the minimum amounts would take up the amount the remaining 40 million Nigerians could not pay. International aid agencies, philanthropists as well as other countries can be approached to symbolically buy some barrels and the entire budgeted income would be met.

Moreover, by 2015 there would be more Nigerians[9] and the burden per person would thus be less. ERA also considers that the local currency would regain strength as corruption goes down and as governance becomes more transparent. If that happens, the naira equivalent of the amount to be contributed by each Nigerian would further decrease. Note that these payments would not need to commence until 2015 and this would give the government sufficient time to take caravans around the nation to explain the beauty of this political and economic move.

Some benefits of keeping the oil in the soil:

- Keeps carbon in the soil thereby tackling climate change
- No oil spills and gas flares from new oilfields
- No destruction of communities or high-sea environments
- No socio-economic ills related to oilfield activities
- Halt to the corrupt nature of the oil blocks allocation exercises
- Illegal bunkering and other forms of oil theft would end since the oil would be left in the ground
- Safe and clean environment
- Reduction and ultimately elimination of violent conflicts in the oilfields.

The proposal considers that the best foot forward for Africa is to halt new oilfield developments and to leave the oil under the ground because Africa cannot afford the luxury of continuing to allow a disconnect between the government and the people, who are forced into the fringes of the formal national economy, accounting for 43 per cent of Africa's GDP (2002).[10] Africa cannot afford to remain in the trap of being a supplier of raw materials at a price externally determined and with environmental costs unattended to.

Decades of oil extraction in Nigeria have translated into billions of dollars that have spelt nothing but misery for the masses of the people. The country offers a model to be avoided and it is time for Africa to step back and review the situation into which she has been plunged. The preservation of the environment, the restoration of polluted streams and lands, the recovery of the peoples' dignity will only come about when citizens stand away from the pull of the barrel of crude oil and understand that the soil is more important to our people than oil and its spoils.

Oil block licensing has become a bazaar in Nigeria.[11] Huge signing fees are exchanged as though the players in the game were soccer or music stars. This signals that there is something fundamentally faulty about the entire enterprise. From audit reports carried out for the Nigerian Extractive Industries Transparency Initiative, even these signing fees are not handled in an open way. In other words, some of them fall into leaky baskets. Nigeria was richer through her great agricultural produce before the ascendancy of crude oil as the nation's major foreign exchange earner. Crude oil brought about crude actions in every realm of national life. It is hoped that if oil is left in the soil Africa would have a better future as its nations looked away from oil while maintaining a stable economic platform from where to leap unto greater heights. This would possibly be through agriculture with supporting governmental structures. Dependence on oil rents must be ended to rescue a largely damaged psyche and build a sense of commitment to nation building.

Come to think of it, if every African were to contribute to their national purse, it would make it clear to politicians that when they misappropriate public funds they are indeed stealing from the suffering people and not from an invisible pile of black gold, diamonds or copper.

The debates and steps to tackle global warming must shift to confronting the root causes of the crises. This demands a quiet appraisal of the socio-economic relations that have birthed the crises. It necessarily requires that humanity rediscovers that it is a part of a cosmos and cannot be bigger than the whole. The current mode of production driven by fossil fuels and other extractives cannot be sustained. Corporate interests drive the hard-nosed push to a cataclysmic point, irrespective of the fact that it is generally agreed that we live on a finite planet. When and why has humanity surrendered its right to live sensibly, accepting instead the corporate creed of greed? What can be done to restore some semblance of balance?

Swimming against the tide, connected by blood

They charged through the mounted troops
They charged through the mounted troops
Sniffing vinegar beneath their scarves
Till the mounted guns dropped and the exalted ones scrammed
They built synergy and spread energy
It was time to connect the tyrants on the maps
Bridge them and abridged their grasps

Time to dust our cardboard armours and tin can caps
Bounce back their plastic bullets & spit in their grumpy faces
We've reached the crucial phase when clanging pots and pans
And flying shoes to boot
Must stand for what we know they should
Time to detach their bloodied fangs from off our bleeding veins

Awoken from our nightmares it's time to dream and to act
They broke the teeth of the blood-sucking vermin to shake off
collective amnesia
Today we see the reasonable thing is demand the unreasonable
If we must recover our memory of proud fighters as we must
And salute the victors
And the living and the fallen in Egypt and Tunisia[1]

RESISTANCE IS ADVOCACY for positive, participatory and inclusive change. The barefaced rape of Africa requires continued resistance in forms appropriate to each circumstance. Without resistance Africa will stay in the pot, like the proverbial frog in the pan, barely noticing the rising heat among so many other survival considerations, until she is cooked in the cauldron.

The thematic focus of Friends of the Earth International (FoEI), an avowedly internationalist federation of autonomous environmental groups, suggests a path that should be adopted and put into practice. Their clarion cry, 'Mobilise, Resist, Transform!' presents us with a template which when contextualised should build into a formidable force for an Africa – and, indeed, a world – where justice is enthroned. We focus here on Africa, but the call is for all citizens of the world. It does not matter how minute or benign the injustices around us may be, every objective situation demands that we mobilise forces, resist those injustices and collectively work and bring about the much-needed transformation.

We keep this in mind as we seek ways to snuff the flames lapping the continent and beat a path to the recovery of the peoples' sovereignty and the enthronement of the best interests of Mother Earth. A central focus beyond building people power is to recapture the commons from the corporates and to de-commodify nature, spaces and relationships.

Post-colonial visions and neo-colonial reactions

There was never a time when Africa was plundered without some form of resistance. Official reactions to resistance continue to follow the same track: criminalisation and annihilation. The historical roots can be found in the pre-colonial scrambles for Africa and in ongoing events. The criminalisation of dissent was the means of cutting down nationalists who sought to pave independent paths that would probably have also heralded economic independence.

This vision was captured in a speech by Samora Machel where he said, 'Our final aim is not to hoist a flag that is different from the Portuguese one, or to hold general elections – more or less honest – in which blacks rather than whites are elected, or to have a black president instead of a white governor.' He went on to declare,

'We affirm that our aim is to win total independence, to establish people's power, to build a new society without exploitation for the benefit of all those who consider themselves Mozambican.'[2]

So many great Africans with this vision are no longer with us. We count, for example, the murder of Congolese Prime Minister Patrice Lumumba in 1961; the liberation-war deaths of Eduardo Mondlane of Mozambique in 1969, Amilcar Cabral of Guinea in 1973, Zimbabwean Herbert Chitepo in 1975, and South Africans Steve Biko in 1974, Ruth First in 1982 and Chris Hani in 1993; the convenient plane crash in South Africa's airspace that took the life of Machel in 1986; the assassination of Thomas Sankara of Burkina Faso in 1987; and Ken Saro-Wiwa's execution in 1995.

A quick review of events that led to the assassination of Lumumba provides further insight into the power of resource grabbing, today labelled energy security, to destabilise nations in the guise of building democracy. When the Belgian colonial power in Congo in 1959 announced a five-year transition programme, nationalists like Lumumba saw that as a ploy to delay the process of independence so as to raise and install puppets who would do the bidding of the colonial masters. Following a rejection of this prolonged transition the Belgian colonialists unleashed terror on the dissenters, killing 30 Congolese during a clash in Stanleyville and sending Lumumba to prison on trumped up charges of fomenting riots.

The Belgians were forced to call elections earlier than they had planned and Lumumba's party, the MNC, won a landslide 90 per cent of the votes in Stanleyville. With the turn of events, the colonial government began to discuss the process of disengagement in earnest. They held a round table conference in Brussels in January 1960 and were forced to release Lumumba from prison to facilitate his participation. It was at this meeting that 30 June 1960 was set as the date for independence after elections. The MNC won a clear victory at the election and Lumumba was invited to form a government as the prime minister, while Joseph Kasavubu took office as president. Patrice Lumumba was in the saddle of power for barely six months before his assassination at the command of a Belgian army officer and despite the protection provided for him by the United Nations as he sought to travel to Stanleyville, an area controlled by his supporters.

Units of the Congolese army opposed to their Belgian commander rebelled and this provided the cover under which the mineral-rich Katanga province proclaimed secession. In the guise of going in to protect Belgian nationals, the colonial army went to Katanga to defend and nurture the secession led by the infamous Moise Tshombe. With the nascent government yet to settle down, a split emerged between the prime minister and the president, and Colonel Joseph Mobutu, leader of the Congolese army, seized power in a military putsch. The path for the plunder of Congo, later named Zaire, was thus laid on the blood stained sands of Katanga. More than other resource-rich countries in Africa, DRC remains peculiarly volatile on account of its rich stock of resources.

In Kenya, the fight for independence had some roots in the resistance of the people to being mere sharecroppers on their own land. The Mau Mau resistance pushed against British land grabbers in the early 1950s. At the height of the struggle, the British brought in troops from England and bombed out the Aberdares and also the Mount Kenya area. In the ensuing confrontations, 100 Britons lost their lives while thousands of Kenyans were either killed in combat or hanged.[3]

Was Africa finally pacified by the elimination or removal from office of the most progressive nationalist leaders in the immediate post-independence years? Is it right to characterise Africa today as being run by bloodthirsty warlords who specialise in using child soldiers and plundering resources in creeks of blood? Is Africa a continent turned against itself and left behind by everyone – except that her minerals resources and, now, land are objects of desire? What we find on closer examination is a web of resistance, one that has not been broken by years of repression. The struggles that emerged took many forms, especially through civil society organising or popular uprisings. We begin, however, with the power of women.

Women's power

When a full history of the struggles for people's liberty is written, the roles played by women will not be ignored. In military terms, the Amazons of Dahomey[4] stand out for bravery in battle. The Amazons were an all-female army that provided bodyguards for the king of Dahomey and also took part in battles. The formation came into being in the 18th century and lasted for over a hundred years. The Amazons were a force of 6,000 to 8,000 women, about a third of the Dahomey army. Similar troops were found among the Ashanti in what is now Ghana. They were fearless and skilled.[5]

Although the role of women in the political development of African society is hardly projected as significant, the truth is that they are dominant in the informal labour force and food production system, as well as strong political actors. The Aba women's riots against unjust taxation in Nigeria (1928–1930) and the more recent resistance of Niger Delta women to the reckless behaviour of oil companies in their territory stand out as vivid examples in which women have been in the vanguard for change.

In combating Chevron's environmental impunity, the women of the Niger Delta utilised the power of their nakedness to occupy and shutdown oil installations from 2002 to 2003. Taking a cue from these women in the Niger Delta, 'Naked protests multiplied around the world as women were inspired to bare all to resist the Bush attack on Iraq.'[6] The naked option was the ultimate weapon that the women wielded against the aggressive oil corporations. Their strategies also included breaking up deals made by men with the corporations which did not address the women's root grievances – such as the blatant destruction of their fishing grounds and farmlands by oil spills and gas flares. The women equally forged alliances with other men who strengthened their demands for the protection of their livelihoods and jobs for their husbands and sons.

In those days of non-violent resistance led by the women, Felicia Itsero, then 67 years old, told field monitors from Environmental Rights Action:

> We are tired of complaining. The Nigerian government and their Chevron have treated us like slaves. Thirty years till now, what do we have to show by Chevron, apart from this big yard and all sorts of machines making noise, what do we have? They have been threatening us that if we make noise, they will stop production and leave our community and we will suffer, as if we have benefited from them. Before the 1970s, when we were here without Chevron, life was natural and sweet, we

were happy. When we go to the rivers for fishing or forest for hunting, we used to catch all sorts of fishes and bush animals. Today, the experience is sad, I am suggesting that they should leave our community completely and never come back again. See, in our community we have girls, small girls from Lagos, Warri, Benin City, Enugu, Imo, Osun and other parts of Nigeria here every day and night running after the white men and staff of Chevron, they are doing prostitution and spreading all sorts of diseases. The story is too long and too sad.[7]

The Saro-Wiwa legacy

The height of non-violent organising in Nigeria was reached in the Ogoni struggle championed by the Movement for the Survival of the Ogoni People (MOSOP) under the leadership of Ken Saro-Wiwa. The Ogoni people organised themselves into MOSOP in August 1990 and quickly produced the foundational Ogoni Bill of Rights.[8] The Ogoni used their Bill of Rights to articulate their grievances with the Nigerian state as well as against the Shell Petroleum Development Company, which was operating in their land. The document also set out the demands of the Ogoni people for economic and political integration into the Nigerian state while maintaining their rights to develop as a people. In addition, they demanded a halt to the destructive oil extraction activities of Shell and the remediation of their despoiled environment. If observers expected the emergence of MOSOP to be a fad that would fizzle out, they were mistaken. Under the charismatic leadership of Ken Saro-Wiwa, MOSOP became a mass movement of the Ogoni and galvanised support from both within and outside Nigeria. The movement also had active wings for young people, women and students. It was a force that could not be ignored.

Although MOSOP preached the gospel of non-violence, the Nigerian state apparatus, then controlled by the military, chose to use force to silence community voices. While the people expected that government would respond positively to their demands, what they got was a rude shock. The military government set up a special military task force to sort out the Ogoni people. This task force is the mother of the Joint Military Task Force – popularly called the JTF – that became a permanent feature in the governance of dissent in the Niger Delta.

The military task force unleashed against the Ogoni people was led by a notorious army officer, Colonel Paul Okuntimo, who, after visiting mayhem on the people, reportedly boasted that he knew over a hundred ways of killing a person and that he had only tried a few so far. This was the era in which the military used their logistical and financial powers to engage in wasting operations in Ogoniland. By 1993 the Ogoni people had excluded Shell from their land and the company hasn't operated there since. The November 1995 hanging of Ken Saro-Wiwa, after a kangaroo court handed down a death verdict on trumped-up charges of involvement with the murder of four Ogoni leaders, intensified the resolve of the Ogoni people, and they prevented Shell from returning to this land and the oil the company so coveted.

At the end of 2009 Shell and the government of Nigeria hired the United Nations Environment Programme (UNEP) to assess the extent of their decades-old pollution in Ogoniland. Controversies erupted over whether the report was fatally compromised by Shell's funding of the process. The controversy was inflamed when a UNEP official was quoted as saying that local people caused about 90 per cent of the oil spills in Ogoniland. In the heat of the controversy, Shell claimed that they doled out the $9mn required for the assessment based on the polluter-pays principle. This admission of complicity stands in contrast to the company's earlier, persistent claims that they are not the polluters.

While some may see the Shell–UNEP partnership as an example of public–private cooperation whereby a public entity is hired to assess the mess made by a private entity, others see it as a perfect fit for a different metaphor: governments as shoeshine boys of corporations. If Shell is right in its assertion that no other oil company operating in Nigeria can do better than they do in terms of social, environmental and other records, then this should permit a very different conclusion: no oil company should operate in the Delta. The Ogoni have a strong case when they insist that their oil must be left underground, as currently championed by civil society groups including the recently formed Ogoni Civil Society Platform and the Ogoni Solidarity Forum.

Community organising has continued in the region. Groups like Chikoko rose to the occasion and worked with the Ijaw Youth Council to champion the Kaiama Declaration in December 1998.

A follow-up Operation Climate Change, in January 1999, aimed at extinguishing gas flares in the region, was met with utmost brutality.[9] The UNEP Environmental Assessment of Ogoniland affirmed that the complaints of the Ogoni people, and by extension other oil-bearing communities, are rooted in fact rather than fiction. In some places the water on which people depend has levels of benzene that is 900 times above World Health Organisation levels. In others the soil is polluted by hydrocarbons to a depth of 5m.

Resistance to Shell, Chevron and the other international oil companies continues unabated, because their destructive extractive activities are as degrading as they have ever been. Yes, they have polished their language, but that is almost about all that has changed in the drilling and killing fields of Africa.

Oilwatch International linkages

One of the major civil society responses was the 1996 founding of a South–South network, Oilwatch International. It took its first steps in Quito, Ecuador and spread its wings across the world, having groups mainly in the South but some in the global North. The signal difference that Oilwatch introduced was that it gave voice and solidarity to community organisations desperately trying to defend their environments. The embryonic stage of Oilwatch was incubated in the offices of Acción Ecológica, an organisation of activists passionate about the health of the Ecuadorian environment, objecting deeply to the environmental pollution in the Ecudarion Oriente and ready to work with communities to demand a change.

Although pragmatic in its demands, Oilwatch remained resolute against destructive fossil-fuel extraction. Time is continuously spent in training community environmental monitors to watch over their environment. The monitors are not academics or professionals carrying out investigations, and realise that their reports may not hold up in a court of law where adversarial judicial systems often require laboratory tests of water, soil and air samples to prove exactly what the pollutants were, when they occurred and who triggered their introduction into the environment.

137

Still, the monitors in the Ecuadorian Amazon – in Lago Agrio, in Jacinto or anywhere else in the Sucumbios – are well trained to know the constituent pollutants in oil spills, wastewaters, drilling mud, gas flares and other toxic outputs of the oil industry. They are trained to take note of changing bio-indicators in their environment. For example, if verdant vegetation begins to turn yellow, wilt and die, they have to take note of that and then observe if there were new human or industrial activities in the area. The monitors are also trained to keep tabs on changing health and demographic patterns.

For Acción Ecológica this was a cardinal approach to raising a brigade of eco-defenders. They used a monitoring camp in the Amazonian jungle just outside Lago Agrio – a camp run in a sustainable way, on solar power and recycled waste water systems. From here they launched monitoring forays on a regular basis.

The springboard of resistance had been laid by massive and barefaced pollution of the Amazon by ChevronTexaco over years of uncontrolled exploitation, unrestrained environmental despoliation and barefaced deception of trusting communities. In summing up the relationship of this corporation with the communities in which they operate, Oilwatch observed:

> The company does not only avoid taking responsibility for their actions but also manages a simple 'solution' to confront the demands of the local societies where they operate. When they are negotiating permits and trying to eliminate resistances, they offer everything States do not provide to their citizens, but when the people complain because of the failure to deliver those offers, and the environmental damage provoked by their operations, due to destruction of property and human rights violations, the company turns around and blames the State. It is a win-win situation for ChevronTexaco.[10]

ChevronTexaco's environmental footprint in Ecuador eventually forced the corporation to quit the country, handing the legacy over to Petroecuador, which, starting from such a heritage, was doomed to get caught in the web of toxic pools and poison. After embarking on a pollution tour of Venezuela, Curacao, Peru and Ecuador in 1997, I was astounded at the massive trail of death left by

ChevronTexaco. It requires impunity and a deadened conscience to do business the way these oil majors operate in the global South. My tour was a part of Oilwatch International's exchange visits, which also saw activists from Latin America visit the Niger Delta in Nigeria. Several exchanges have been conducted between African countries such as Angola, Cameroon, Chad, Gabon, Ghana, Mali, Mozambique, Nigeria and South Africa.

The despoliation of the Ecuadorian Amazon by ChevronTexaco upstages even Shell in Nigeria. Besides gas flares, toxic waste pits and other poisons dumped into the Amazon, ChevronTexaco regularly set ponds of crude oil on fire, causing a crude oil rain that coated everything in grime – roof tops, clothes drying on lines, crops and livestock. The thing about crude oil pollution is that it cannot be wished away. It simply does not disappear, as can be seen in Ecuador several years after Chevron departed.

Finally, justice was announced in 2011: an $8.6bn fine on Chevron for heavily polluting the Ecuadoran Amazon through serial oil spills between 1964 and 1990 . Chevron bought Texaco in 2001 and so inherited its liabilities.

The case against Chevron was first fought in a New York court, when the plaintiffs sought $27bn in a case first filed in 1993. But, on the expectation that it would be easier to intimidate a third world judge, the firm succeeded in getting the court to agree that the legitimate place to try the case was Ecuador. The plaintiffs and their young legal team were, however, tenacious. In a statement issued soon after the judgement was delivered, Chevron declared, 'The Ecuadorian court's judgment is illegitimate and unenforceable,' and that it was 'the product of fraud and is contrary to the legitimate scientific evidence.' Their statement was laced with open threats: 'Chevron does not believe that today's judgment is enforceable in any court that observes the rule of law … Chevron intends to see that the perpetrators of this fraud are held accountable for their misconduct.' And, of course, Chevron went on to appeal against the judgment. Because Chevron refused to accept the judgement and apologise, the fine soon doubled.

Corporations such as Chevron fight hard to avoid liability even when caught in the act with a smoking gun still in their hands. The people of the region suffer the impacts of the pollution on their health through diverse cancers, blood disorders and other such diseases. The corporation also left behind pipelines that often run above the ground, sometimes so that people have to stoop beneath them to get into their homes. One of the more atrocious acts of 'corporate social responsibility' was the practice of using toxic drilling muds to make building blocks for schools in the area. Reports also abound of toxic wastes from oil activities being spread on community roads as a social service. The sad thing, however, is that after the operations were taken over by the national oil company, Petroecuador, there still does not appear to be significant respect for the environment or the people in the region.

Chevron perhaps sought to distract attention from the verdict in Ecuador by moving across the Atlantic, to the bloodstained and oil-soaked creeks of the Niger Delta. The link and the timing were inescapable. The company announced with much fanfare a splash of $50mn, ostensibly to ignite economic development and tackle conflict in the region – to which, it must be said, the company admitted to being a contributor in the past. They aimed to channel funds through the company's Niger Delta Partnership Initiative and the United States Agency for International Development (USAID). But this is tokenistic, since the oil company hires only a tiny fraction of the millions of people it has impoverished through the destruction of the creeks, swamps, farmlands and forests that they depend on for their livelihoods. A few million spent on community development cannot compensate for the oil spills, gas flares and the dumping of other toxic wastes.

The company has loads of money and plenty of time. The poor indigenous people and the *campesina* in the Amazon forest or the Niger Delta have neither of those. To the poor people it is a fight for survival. For the oil mogul it is an image thing. When Chevron bought Texaco, it should have been obvious that they purchased both the assets and the liabilities.

From the Amazon to Ilaje

ChevronTexaco stands accused of atrocious violations of human rights in Ilaje communities of the Niger Delta. In one incident, on 28 May 1998 at Parabe oil platform, Chevron's assaults involved the use of the Nigerian military aboard helicopters provided by Chevron. It was a case of summary executions of unarmed youths, torture, and wanton destruction of lives and properties. A case ensued in United States District Court in San Francisco, and although the jury initially ruled against the plaintiffs in Bowoto et al v Chevron, the context certainly demonstrates the blatant socio-economic, ecological and political injustice that pervades this area.

Up to 20 per cent of ChevronTexaco's total oil production in Nigeria comes from the Ilaje communities. The firm has operated here as Chevron Nigeria Limited (CNL) since 1962 when it started exploratory activities, and in 1968 oil exploitation began at the Okan Oilfield. Other oilfields such as Meren, Parabe, Isan, Malu, Ewan, Opollo and Opuekaba were added from 1968 onwards.

The oil giant's activities in the area raised hopes and dashed them immediately, as inevitably happens in oil-dependent countries of the South. By the time of the 1998 incident, after 30 years of uninterrupted crude oil exploitation, the people of Ilaje had been given only the following benefits from the oil giant: a single wooden secondary school for the 42 communities which stretch over 60km along the Atlantic coast, two jetties and one bore hole (that was out of commission). In the same period, the oil company had employed 2,500 Nigerians, but only two people from the surrounding community.

One Chevron official who played a role in the debacle stated in his 2003 declaration to the court, 'CNL and its contractors tried to hire as many community people as possible and often also hired "ghost workers" such as community members who were not needed for a project or who lack the relevant skills. These are also members who were put on the payroll even though they were not expected to report for work.' He also stated that in order to make these ingenious arrangements they had to identify the communities' representatives, and that it was extremely difficult 'for us at CNL to distinguish legitimate representatives of the host communities from individuals pursuing their own schemes of extortion or fraud'.

By this arrangement, Chevron encouraged fraud by doling out salaries to persons who did no work, but served as a standby team of reservists who could be relied upon to defend company interests even against those of the community. Their method of identifying community leaders was arbitrary as they apparently did not use extant community structures and instead erected parallel ones. These are all methods of dividing communities and building tensions and conflicts.

When I visited the territory ten years after the Parabe platform killings, things had degenerated even further. The canals opened by Chevron for their equipment continue to degrade their fresh water sources due to the intrusion of salt water from the open sea. On account of the destruction of fresh water sources, some Ilaje people have to depend on wastewater from a borehole built by Chevron at Opuekeba. The borehole at Opuekeba was made for the purpose of cooling Chevron's gas turbine and the people fetch the hot water that comes from this installation and must wait a long time for the water to cool and for sediments to settle before they can use it.

Community people had approached the company facility, merely asking if they could talk to Chevron over their environmental and livelihood concerns. The community made many efforts to meet with Chevron and also sought help from government channels to make this happen. When these efforts failed, youths decided to embark on a peaceful direct action. On 25 May 1998, 121 of them landed on a barge attached to the Parabe platform. The following day, Deji Haastrup, CNL's community relations manager, met with the youths on the platform. The youths demanded that senior Chevron officials meet with leaders in the Ikorigho community. At the meeting the next day, the community asked CNL for the following:

1 Restoration of their environment to its natural state.
2 Employment for community youths.
3 Location of an operational office or tank farm in Ilaje land.
4 Construction of an embankment to check sea incursion and coastal erosion.
5 Provision of social amenities such as a model school, potable water supply, and a town hall as Chevron often complained of a lack of meeting venue in the community.
6 Negotiation for any work to be done in Ilaje land.
7 Compensation for loss of fishing grounds to the elders.
8 More scholarships for the youths.
9 Appointment of indigenous Ilaje contractors as permanent contractors, as was the practice in other states of Nigeria.

After the meeting the officer asked for time to consult with his head office in Lagos and said he would send a response two days later, on 29 May 1998. On receiving this information about the progress of the talks, the youths on the barge decided that they would discontinue their protest the following day and return to Ikorigho to be a part of the negotiations. But before they could leave, soldiers landed on the Parabe platform at 6.30am in three Chevron helicopters. The soldiers reportedly shot at the youths indiscriminately and the unarmed youths simply had no chance. Two were killed and 30 suffered gunshot wounds. The fact that the youths were not armed was confirmed by Chevron's own declaration.

Survivors and representatives of the communities filed the suit Larry Bowoto et al v Chevron. The plaintiffs included Larry Bowoto, Bola Oyibo (who died in 2001), Bassey Jeje and Sunday Johnbul. The case took eight and a half years of litigation and four weeks of trial and demanded intensive and committed work by the plaintiffs and their lawyers (Traber & Voorhees, as well as the private law firms of Hadsell & Stormer and Siegel & Yee, the Center for Constitutional Rights and the Electronic Frontier Foundation). Attempts by Chevron to have the case thrown out of the US court or tried in Nigeria were rebuffed by the judge, who agreed with the prosecutors that there was an undeniable umbilical cord linking Chevron to the incident. When the case eventually went to court, all the plaintiffs other than Larry Bowoto and scores of witnesses were travelling out of their immediate communities for the first time. Most spoke no languages other than Yoruba and Nigerian English.

When judgement was delivered on 1 December 2008, the jury cleared Chevron of all charges. This was in spite of the ruling by Judge Susan Illston of the San Francisco district court that CNL 'personnel were directly involved in the attacks; transported the Nigerian security forces, paid the security forces and knew that the security forces were prone to use of excessive force'. Richard Herz, an attorney at EarthRights International and co-counsel for the plaintiffs, reacted to the verdict saying, 'The entire history of crude oil extraction in Nigeria has been strewn with dollars for the corporation and politicians but tears and blood for local people. Chevron must account for its unrelenting acts of inane brutality to the defenceless people of the Niger Delta region ... Although the plaintiffs did not prevail, Chevron now knows that it cannot conceal complicity in human rights abuses from public scrutiny.'[11]

The outcome of the case has not shut down the judicial quest for justice. After the ruling, the arrogant Chevron attorneys sought compensation from the communities when they filed a bill of costs for $485,159. However, the judge agreed with the plaintiffs that the factors weigh against awarding costs to defendants. The judge found that:

> The economic disparity between plaintiffs, who are Nigerian villagers, and defendants, international oil companies cannot be more stark. The trial plaintiffs have filed declarations attesting that they cannot afford to pay any costs or fees in this case. They have jobs working in a petrol station (earning as much as $100 per month), operating a kerosene business ($867 per month), doing odd jobs such as cutting or selling firewood, fishing, and construction ($60 per month); selling spare parts for fishing boats ($1271 per month); buying and selling fish ($465 per month); collecting rent ($207 per month); buying and reselling clothes ($567 per month); operating a small store ($233 per month); or are students – ten of the plaintiffs are minors – and have no income.[12]

The court also found that awarding costs to defendants in this case would have a 'chilling effect ... on future civil rights litigants'. The court declared that:

> At root, this case was an attempt by impoverished citizens of Nigeria to increase accountability for the activities of American companies in their country. Plaintiffs' ultimate failure at trial does not detract from the fact that this was a civil rights case. The threat of deterring future litigants from prosecuting human rights claims in the future is especially present in a case such as this, where plaintiffs have paltry resources and defendants are large and powerful economic actors. Relatedly, the Court finds that this case presented close and complex questions. Nothing about the case was simple – from determining the applicable law to obtaining permission for plaintiffs to travel abroad, this case presented unique challenges.[13]

When Chevron claimed that the plaintiffs' failure at trial demonstrated that their claims never had merit to begin with, the court disagreed with Chevron.[14] The plaintiffs filed an unsuccessful appeal against Chevron, seeking a reversal or retrial. In the appeal, the attorneys for the plaintiffs observed that Chevron claimed that the Ilajes' mindset was wired to favour 'hostage-taking' and 'kidnapping' and that 'it's something that they would do'. The appeal also stated that the District Court ought to have placed the burden on Chevron to prove that they were acting in self-defence when they orchestrated the killing of unarmed protesters at the Parabe Platform.

In spite of the initial defeat, the case added strength to the web of resistance against reckless oil majors. The web will get stronger and larger. Inspiring struggles for environmental justice pop up in many parts of the world and there is an urgent need to weave these together into a global force to liberate Mother Earth from the claws of miners and speculators.

Take the example of the August/September 2011 protests at the gates of the White House in the US against a tar sands pipeline proposed to link the mines in Canada to refineries in the US. The protests quickly echoed around the world as activists joined in solidarity in Brazil, Egypt, Germany, India, Peru and South Africa. The internationalist nature of environmental justice protests points the way to redirecting power relations in a world with values skewed against nature and against the less powerful.

The protests at the White House were met with the arrest of hundreds of citizens, but the activists had taken a determined stand in civil disobedience and had drawn a line in the sand. Protesters are concerned about the catastrophic impacts of tar sands on the climate as well as the impacts of the pipeline and related toxic substances on water resources and wildlife. There have been reports of a rise of rare cancers among First nation peoples of Canada who live close to the tar sands fields.

The proposed Keystone XL pipeline is designed to transport 700,000 barrels of crude oil per day to delivery points in Oklahoma and south-eastern Texas. The 36-inch pipeline would consist of about 327 miles of pipeline in Canada and 1,384 miles in the US. According to Tom Goldtooth, executive director of the Indigenous Environmental Network:

Our Indigenous-Native Nations of the U.S. and Canada must unite to oppose the Keystone XL pipeline and come together to find local, clean, renewable energy to reduce our carbon footprint and spur the economy. There are too many major safety, environmental and public health hazards possible in the Keystone XL Pipeline project. The cost and risks of building an oil pipeline across our traditional homelands with important aquifers, waterways, natural lands and wetlands is too great at this time. Our homelands within the planned corridor of this pipeline have many cultural and historically significant areas that have not thoroughly been assessed and are in danger of being destroyed. The negative and very destructive human rights impacts of the Keystone XL pipeline transporting dirty oil from the tar sands region of northern Canada have not adequately been assessed in the final EIS. First Nations in the tar sands region have consistently been making reports of devastation of their environment, their waters, air, and more recently their health. Cancers linked to petroleum contamination are on the increase.[15]

The ideal of community ownership of community resources requires that we understand that this means the people actually have full rights over their lands and territory. It also supposes that people have the right to free, prior, informed consent before any action or activity is executed on their land or territory. It precludes situations where a handful of leaders claim to represent communities and become pawns in the hands of capital, sell their people's patrimony and load their pockets with the loot.

In contrast, the maxim in today's global political landscape is that might is right. The rise of unilateralism under the US presidency of George W. Bush rendered multilateralism more or less cosmetic. The rise of prescriptive neoliberalism has shattered state sovereignty. And couching all this in terms of advancing the democratic ideals of liberty and fair competition allows Washington and its allies to fib about military humanism. This scenario has been appropriately captured by Naomi Klein as 'disaster capitalism'[16] – a situation where disasters are seen as opportunities to impose a pre-planned superstructure that inevitably denies powerless citizens of the world their rights. The whole idea is to hit the people so hard that they are pushed into a state of shock and while in that condition they are unable to react collectively or cogently to the harm being inflicted on them. Such disasters are increasingly man-made, although even natural disasters are equally exploited to dispossess the weak. An example can be seen in opportunistic slum or seafront clearances where the land is then grabbed by big ventures.

The path of crude oil development has been strewn with skeletons and soaked in human blood across the world. The ongoing case in Nigeria is a glaring example. Memories of a three-decade war in Angola – beginning in 1975 and arguably fought over oil – are still fresh. In 1999, as the first barrels of crude oil were shipped from Sudan, so did the war between government forces and those of the Sudanese People's Liberation Army escalate. When we turn our eyes to the Middle East we see the raw situation of war waged for profit and resource appropriation and control.[17] If this scenario should blossom unchecked, what we experience today will end up being nothing more than a whimper. There appears to be a direct correlation between oil and violence of various kinds. How best to fight back, so as to weave yet more threads into the web of resistance and transformation?

Recall that the MOSOP struggle over Ogoniland showed us an excellent example of non-violent struggle for community rights, and recall the violent response of the state, a signal that dissent would not be tolerated. When other parts of the Niger Delta erupted in violent confrontations between 2005 and 2009, the ultimate response before the Nigerian president offered an amnesty was a military assault on Gbaramatu kingdom and neighbouring communities. Armed militants rose in the oilfields for a plethora of reasons. For a couple of years kidnapping and hostage-taking became very lucrative business for some of the so-called militants. Oil company workers and the military personnel found good business in negotiating the release of hostages as well as in the securing of contracts with militants for the protection of oil pipelines and other installations.

The levelling of Gbaramatu community in May 2009 was achieved by a combination of aerial and sea raids, an object lesson to the warlords at militant 'Camp 5' located close to the community. When I visited one of the refugee camps at Ogbe Ijoh, near Warri, one of the survivors, Mrs Akpoaboere Helpme of Okorekoko kingdom, recounted that when the invasion took place the thundering sounds of gunshots came unexpectedly, with people scampering helter-skelter with little or no thoughts and time to rescue their children and the elderly. In her own bid to save her mother and baby, her six weeks old daughter fell in the river and in her panic state of mind she did not realise this for a while until she saw the baby floating miraculously on the water.[18]

Soon after this the Nigerian president announced an amnesty that was to expire in October 2009 and offered to rehabilitate those who laid down their arms. One after the other, the warlords scampered around for handshakes of the physical as well as financial variety. Some of the less prominent warlords were displeased at not receiving the same sort of recognition that others did. Some complained that while their colleagues were hosted in five star hotels they had to be content with sleeping under the zillion stars in the sky.

The ending of militancy through an amnesty declaration raised a number of questions. One of the well-known leaders of armed groups, Alhaji Asari Dokubo, rejected an offer of amnesty, saying that he had never committed any crime and that if anything he had engaged in a just struggle for self-determination. Others, such as Henry Okah, embraced the amnesty but later sent signals that suggested the stunt might not end militancy in a sustainable way. Okah said in an interview on the cable television channel Al Jazeera that the Nigerian government had exaggerated the gains of the amnesty and that the peace achieved was tenuous indeed. He was also quoted as saying that the actual number of militants who had surrendered weapons in the amnesty exercise was not even 4 per cent of the numbers claimed by the government.[19]

In fact, two leaders of armed groups, Chief Government Tompolo (whose Camp 5 near Gbaramatu was crushed by the military) and Ateke Tom, at a point declared that they regretted accepting the amnesty. Their grouse was that a month after the amnesty was accepted all they were engaged in was an endless series of meetings with officials from the presidency without a clear indication of what the outcomes would be. According to Chief Government Tompolo, 'There was no need for endless meetings ... things should be happening naturally and very fast in the Niger Delta after the amnesty programme. We have discussed issues of development of the region over and over. We don't need any expert or consultant to come and tell us what we need in the region.'[20]

Despite misgivings on the construction of the amnesty programme by the Nigerian government, a level of peace has prevailed in the oilfields as evidenced by the fact that by early 2011 the country was able to attain its OPEC oil production quota. At the height of militant activities in the region, oil production had fallen sharply.

Analysts of the origins of armed resistance in the Niger Delta have claimed that it started out as political thugs hired by the ruling political party in Rivers state for the purpose of election rigging. Other groups are said to have risen for the political reason of seeking justice for a viciously exploited region. Some who specialised in kidnapping parents, wives or children of politicians and other persons perceived to be rich, for the purpose of receiving a ransom, may have engaged in nothing other than plain criminal activities. But there were others who claimed that men who were engaged in the business of crude oil theft, locally called illegal bunkering, also raised armed groups.

Speaking at the National Institute for Policy Studies – a respected institution for top policymakers in Nigeria – the governor of Nigeria's Delta state, Emmanuel Uduaghan, asserted that oil companies were involved in the illegal activities. He also said that the international community was complicit in the thefts since there was a ready market for the stolen crude.[21] In the view of Dimeji Bankole, at that time speaker of the Nigerian House of Representatives, about half of Nigeria's crude oil production is stolen. He summed up that if that estimate is correct, it means that Nigerian crude may run out sooner than expected.[22] Other analysts believe that Nigeria is actually losing more of the crude oil than it is selling officially as a result of the connivance of security agencies that are meant to halt such practices. Through this the country is estimated to lose US$1.6bn to oil thieves annually. Reports allege that some top naval officers, serving and retired, have private pipelines that run from the Port Harcourt area to Eket and that these pipes serve as conduits through which they siphon crude oil, which is loaded onto vessels and shipped to refineries on other shores including South Africa.[23] If this is true, then the international community is guilty of abetting and fuelling the collective robbery of the people of the Niger Delta of their patrimony. As for the involvement of security officials in this perfidy, it is a sad commentary on the connivance of African leaders in the pillage of the continent.

As though it was a season of amnesties, the president of Niger also declared an amnesty for Tuareg rebels who were fighting for a higher slice of the revenues accruing from oil and uranium mining in the northeast region of Agadez. The rebels, including the main group, the Niger Justice Movement, laid down their weapons in June 2009 after two years of fighting.[24]

The destruction of the Niger Delta community of Gbaramatu and nearby communities before the amnesty to the militants was not the first time that the Nigerian state had descended on a resource rich-community with maximum force. Interestingly, oil companies evacuated their personnel from the area shortly before the bombardment started, suggesting that they may have received warnings that an assault was about to take place in the area.

When Tombia town was attacked by the Nigerian military ostensibly to dislodge armed fighters, Shell reportedly evacuated its staff with helicopters just before the attack. As captured in *Crude World*,[25] 'helicopters and speedboats all but destroyed it [the community], though soldiers, before torching the better huts, made of wood, looted them of anything that merited looting, including pots and pans.' Two villagers were killed in that attack.

The deepest conflict and violence have not been inflicted by the power of the gun, nor by the burning of Umuechem;[26] the massacre at Odi[27] or Odioma; the devastation of Ogoni and the extra-judicial murders of Ken Saro-Wiwa and other leaders. The greatest violence is the mass poisoning of communities through pollution.

Genuine national leadership

Late November 2009, while visiting the Nigerian minister of information, the Venezuelan ambassador, in a suspension of diplomatic niceties, spoke from his heart about what he felt about his host nation. It was particularly striking as the Nigerian minister of information's pet project was the rebranding of Nigeria, a project that some observers equated to using colourful bandages to make a festering sore appear acceptable. This is what ambassador Enrique Fernando Arrundell told the minister:

In Venezuela, since 1999, we've never had a rise in fuel price. We only pay $1.02 to fill the tank. What I pay for with N12,000 here (Nigeria), in Venezuela I'll pay N400. What is happening is simple. Our President (Hugo Chavez) decided one day to control the industry, because it belongs to the Venezuelans. If you don't control the industry, your development will be in the hands of the foreigners.

You have to have your own country. The oil is your country's. Sorry I am telling you this. I am giving you the experience of Venezuela. We have 12 refineries in the United States, 18,000 gas stations in the West Coast. All we are doing is in the hands of the Venezuelans.

Before 1999, we had three or four foreign companies working with us. That time they were taking 80 per cent, and giving us 20. Now, we have 90 per cent, and giving them 10. But now, we have 22 countries working with us in that condition.

It is the Venezuelan condition. You know why? It is because 60 per cent of the income goes to social programmes. That's why we have 22,000 medical doctors assisting the people in the community. The people don't go to the hospital; doctors go to their houses. This is because the money is handled by the Venezuelans. How come Nigeria that has more technical manpower than Venezuela, with 150 million people, and very intellectual people all around, not been able to get it right? The question is: If you are not handling your resources, how are you going to handle the country?

So, it is important that Nigeria takes control of her resources. We have no illiterate people. We have over 17 new universities totally free. I graduated from the university without paying one cent, and take three meals every day, because we have the resources. We want the resources of the Nigerian people for the Nigerians. It is enough! It is enough, Minister![28]

It is enough, Minister! It is enough, President! These words could have been spoken to any minister of any of the resource-rich countries of Africa wallowing in the self-delusion of greatness while kicking around in the mire of corruption and wastefulness. The Venezuelan chastised the Nigerian official for wasteful utilisation of funds from crude oil. We take this as a metaphor for wasteful use of funds from any natural resource and keep in mind that the best value of fossil fuels is to leave them in the soil. Revenue derived from crude oil exploitation, for example, can hardly finance restoration efforts that may be needed following years of impacts on the environment and peoples.

Great African leaders like Kwame Nkrumah, Thomas Sankara, Patrice Lumumba, Amilcar Cabral, Samora Machel, Julius Nyerere and others would be shuddering in their graves if they were to witness the plunder that has ripped Africa to bits thanks to the connivance of leaders whom transnational corporations, venture philanthropists and international financial advisers have led by the nose. Sankara illustrated the African spirit needed to realign the continent away from economic and political poverty and towards liberating ideas and peoples' sovereignty. Some may think Sankara was an idealist and thus left his flank open to deadly bullets from guns wielded by friends. But though these leaders are dead, their ideas cannot be killed – as Sankara himself declared a week before he was assassinated.[29]

The ideas that Sankara espoused rattled many of the corrupt presidents and ministers on the continent; they are still potent today and can only be ignored to the continent's sorrow. He was engaged in a struggle against corruption, long before it became the hypocritical song on the lips of the World Bank and the IMF. He realised quite early that corruption was used as a tool by the international capitalist mafia to conquer markets and pillage the resources of the global South.[30] Talking about the debt trap, he could not be faulted when he declared, 'If we do not pay the debt, our lenders will not die. However, if we do pay it, we will die...' He demanded the repudiation of all odious debt, as nothing short of modern-day slavery.

Sankara equally realised that the mobilisation of the people was vital for the emancipation of the continent from the clutches of destructive extractors. To him, the people held the key to the struggle, and they fully understood both the glory to be gained and the pains to be borne in the process of regaining control over their lives, environment, resources and destiny. His conviction was that there was no sense in speaking on behalf of the people, but rather to stand in solidarity with them, have them integrated into the struggle and, as Sean Jacobs describes:

> develop an identity forged in the fire of action. For Sankara: 'I think the most important thing is to bring the people to a point

where they have self-confidence, and understand that they can, at last … be the authors of their own well-being … And at the same time, have a sense of the price to be paid for that well-being.' To a great extent, the Burkinabé Revolution was an original experiment in profound social, economic, political and ideological transformation. It was a bold attempt at endogenous development through popular mobilisation.[31]

Permit me to linger a bit with the memory of Sankara and the reforms he instituted after coming to office in 1984:[32]

- He changed the country's name from Upper Volta to Burkina Faso – the land of the upright people.
- He believed in economic self-reliance, promoted local food and textile production and refused World Bank loans. His targets were policies concerned with the real needs of the people and not the dictates of the World Bank or the IMF.
- He embarked on a land reform whereby the land belonged to those who cultivated it.
- He was one of the early leaders to be seriously concerned with environmental protection – especially against desertification.
- Sankara banned the system that allowed tribute payments and obligatory labour to village chiefs, abolished rural poll taxes, promoted gender equality in a very male-dominated society (including outlawing female circumcision and polygamy), instituted a massive immunisation programme, built railways and kick-started public housing construction. His administration aggressively pushed literacy programmes, tackled river blindness and embarked on an anti-corruption drive in the civil service.
- He declined to have his photos displayed in public buildings, in stark contradistinction to those who usurped power after his murder.
- Sankara earned a small salary (about $450 a month) and eschewed luxuries, including first-class flight tickets.
- He fought corruption and carried out administrative reforms for good governance.
- He believed in women's emancipation and participation in various facets of national life.

Many African governments operate as though the people do not exist except as objects to be exploited or as subjects to be suppressed. This happens because they do not derive their mandate or legitimacy from the people. They remain in office according to their personal pleasure. This is never a problem for the extractive industries, which are happy to reinforce the thrones of dictators as long as it provides them with a stable atmosphere for unmitigated exploitation of the territories. This is the tragedy of Africa. The appeal of Sankara is that his people were his forte and his vision was to make them stand proud and with integrity.

Experience in the field shows that resistance to destructive extraction has to be built one block at a time. When all blocks link together, there forms a wall – sometimes protecting a whole nation – to block the tide of rapacious exploitation.

Nigerian community politics

28 November 2009: on the road to Goi, in Gokana local government area of Rivers state, Nigeria. I went with a group of community organisers on a community exchange visit and we had to make several detours to avoid the huge craters that have bedevilled the major highway that traverses Ogoniland. The community organisers came from Akwa Ibom state, Edo state, Delta state, Bayelsa state and Rivers state under the umbrella of the Host Communities Network (HoCoN). The core reason for coming to Goi was to see the impact of oil spills from Shell's facilities that had erupted in 2004. They would see the sites and then retreat for reflection. They would see how what took place in Goi approximates with what they have experienced in their own backyards. They would see that in the oilfields of the Niger Delta, pain has been disproportionately weighted against the wretched of the earth.

Goi was a dream community before the black gold began to spurt and then burst into flames. They had fresh water swamps that would get flooded at high tide. When the tide receded the community people had large numbers of fishes to simply pick up or harvest. Others simply built fish ponds by the edge of the swamp and would have them filled up with fingerlings, juveniles and an assortment of fishes. The lush mangroves provided spawning grounds for many varieties.

The story changed once fires from Shell's spills licked the swamp in 2004. Today the swamp is no more; a better description for it might be a crude toxic lake. The tides still flow and ebb, but bring nothing but sad memories to the people whose boats and fishing gear went up in the oily infernos. Where the treasures of the sea used to wash up on these shores, today the people have to wade through the oil slick in search of fresher waters for whatever catch they can get. While we were there, a little lad came back from an overnight foray with a plastic bowl of tiny crabs and some crayfish. For all his effort, his catch was worth a mere 400 naira, if we were to accept his price. The plastic bowl was coated with crude oil and the crayfish and crabs reeked of the stuff too. He would wash the oil off the crayfish before eating them. But that would not detoxify the already poisoned creatures. Moving away from the desolate banks of the swamp with its burnt mangroves and a few surviving trees with oil-coated roots, we climbed away wondering what it must mean to live in this community with the completely devastated environment.

The HoCoN members exchanged tales of woe. The also exchanged ideas on how they could better protect their environment and resist destructive extraction. One resounding conclusion was that for a woman, the best time to divorce a man is before he gets you pregnant. That wisdom was picked up from the Warao community women in the Orinoco Delta, Venezuela, when they were resisting the entrance of BP into their creeks in the late 1990s. They firmly believed that all peoples should stand together in rejection of new oilfield development and help wean the world off the destruction that emanates from the black gold, whose true place should be where nature has wisely buried it.

The 'host communities' concept is one that has been used by both oil corporations and their partner governments to pitch one community against another. In mineral-rich areas, communities in which extractive industry facilities are located are known as host communities while those on whose lands no such facilities are sited are, for example, called non-oil bearing communities. HoCoN redefines the host community as one which hosts industry facilities, or suffers, or would be likely to suffer, impacts of destructive extraction. This broad and inclusive definition removes the sharp daggers used by the extractive industry actors, and unscrupulous government agents, to divide communities and pitch them against one another. Africa needs creative networks such as these in order to build the force needed to unite her peoples.

In the wake of the amnesty granted the militants who militated against peace in the Niger Delta oilfields, the federal government of Nigeria promised, in a statement issued in November 2009, to allocate 10 per cent of oil revenues directly to the communities.[33] This pronouncement raised a lot of interest that soon turned into a discussion about what constitutes a community. The idea of a host community, and its exclusive definition as being the one on whose land an oil company decided to locate their equipment, was loudly trumpeted. But other community campaigners raised the concept that fits the HoCoN view. An example is the view expressed by Oronto Douglas:

> The host community concept is divisive, dangerous, corrupting and has never worked. There is no 'host community' that has become developed in the last fifty years since the discovery of oil. We must resist the 'host' concept and we must encourage our government to reject it as well. As new policies emerge to address our historic grievances, all red flags of combustible characteristics must be identified and neutralised. The community to me is the Clan or the Nationality. Thus, Ogbia, Nembe, Gbaramatu, Kolokuma, Okrika, Kalabari, Ilaje, Mein, Okpe, Ogba, Egbema, Itsekiri, Urhobo, Ibibio, Ogoni, Ikwerre, Etche, Ijaw, Oron, Isoko etc are the communities.
>
> In our very first step towards the control of our resources, we must be clear, deliberate and inclusive. We must not give in to internal division. Where oil and gas is produced or where it is not must not be allowed to divide us … The Ogoni struggle was not a village affair. It was a nationality commitment by all the people of Ogoni for survival and justice.[34]

Notwithstanding the above eloquent plea for an inclusive understanding of the community, others questioned the parameters of the offer by the Nigerian government. For example, the Conference of Ethnic Nationalities of the Niger Delta (CEEND) called for interpretation of the 10 per cent equity for oil-producing communities as well as details of where related capital projects would be sited. Due to lack of confidence in the government, the organisation demanded to know what the offer was based on. '10 per cent equity on what? Is it 10 per cent equity on the joint venture agreement? We want to know the 10 per cent equity shareholding. When we know exactly the details, we will be in a better position to comment on the issue.'[35]

With so much disenchantment with government and transnational corporations, the people must forge alliances between themselves and their communities. It means the reconstruction of solidarity and reclaiming people's sovereignty in all spheres of endeavour. African governments are unwilling or unable to regulate the extractive industries. The industries hide behind the shield of state instruments of repression and get away with impunity. Communities are criminalised when they protest despoliation. The answer lies in linking communities and peoples, sharing ideas in all ways of creative communication, learning from happenings and from history, and being ready to confront the wielders of power. That is the way to reclaim the heavily polluted and overrun community environments. It means setting up community schools on sustainability and environmental justice. It means documenting experience and sharing knowledge and wisdom. It means getting involved in political processes and insisting that leaders emerge from the ballot and not by the bullet.

Solidarity for recovery

Community organising and solidarity building across national and even ethnic boundaries is essential for the recovery of Africa. Communities need to be empowered to regain control over the continent's rich natural resources. Some argue that communities do not have the competence to know what to do with the resources. We would counter by demanding respect for the memory and knowledge of the people. No one can claim to know the environment of any community better than the inhabitants of that community. Communities have over the years learned to live harmoniously with their environments, utilising the resources they are endowed with in a sustainable way. The notion of sacred and sometimes evil forests helped preserve forests and certain tree and animal species. Forests were not just places were timber was harvested or where carbon was sunk so that someone obtained financial rewards or credits. Forests provided a matrix of goods and held a stock of riches for any community.

Social norms preserved both the dignity of the people as well as the health of certain resources. For example, certain parts of streams and creeks were preserved for potable water only and no one would swim or bathe there. Such norms ensured that these water bodies were not polluted. Some communities had certain animals as their totems. While the next community would hunt and decimate that species, they would be conserved in a community where killing them was taboo. Mineral resources were mined and utilised by guilds in such a manner as to preserve the integrity of the environment. The accumulation of surplus or profit maximisation did not overtake the critical need to stay in harmony with the environment.

Community control of community resources would demand that anyone who wishes to exploit such resources would first have to engage in dialogue with the community. In much of Africa today, governments have assumed ownership of all valuable resources including land. They claim to hold such resources in trust for the people, whereas they actually hold them for the satis- faction of the needs of transnational corporations, political leaders and local cronies. Dialogue with communities could lead to the preparation of feasibility studies and environmental, economic and social impact assessments where communities wish to proceed with proposals for exploitation of available resources. Critically, dialogue with communities would be held with the understanding that the community reserves the right to say no to the extraction of any resource in their territory. They must retain the right to decide who should carry out any extractive activity on their land and how it would be done. They should also have the right to decide how the benefits accruing from such extraction would be utilised. Many communities would be willing to release as much as the government wanted in the form of taxes, but the fact of ownership must be vested in the owners of the property, the community.

A word of caution is appropriate at this point. We must not run away with romantic views of communities. Simply because a decision is taken by a 'community' does not confer on it an air of infallibility. When communities take decisions with regard to their resources it must be on the basis of knowledge, although this does not have to be at the level of rocket science. Often we hear officialdom pleading that policies are based on scientific facts that fly above the heads of ordinary people. The truth is that science must be at the service of the community and must be comprehensible. Indeed, when we look at what science has said about climate change and its recommendations over the need for binding emissions cuts to avoid catastrophic temperature rises, we must wonder about governments' and global institutions' studied avoidance of science in that regard.

By presenting partial information, carbon speculators have presented forests as mere carbon stocks to be traded and have succeeded in getting some communities to cede their forests for REDD in exchange for immediate gratification through cash receipts. This can be characterised as outright subterfuge. The displacement of forest communities, the transference of deforestation to other regions, the destruction of local livelihoods and the false promise of fighting climate change does not stop carbon traders from dreaming up more avenues for land, soil and water grabs, not to mention sky grabs. Communities must therefore remain vigilant to ensure they do not become agents for the playing out of a new scramble for Africa.

We mention one example of a land grab where just one 'paramount chief' signed off 600,000 hectares of community land, with a possibility of ceding a further 400,000 hectares. The deal was defeated through resistance by the people with solidarity actions from groups such as the Oakland Institute.

Nile Trading and Development (NTD), a Dallas, Texas-based firm, had entered a deal with Mukaya Payam Cooperative in 2008 in Southern Sudan. Through this deal the company would enjoy a 49-year lease of 600,000 hectares of land at the insulting sum of US$25,000. The lease gave the company full rights to exploit all natural resources in the leased land, including the right to:

- Develop, produce and exploit timber/forestry resources on the leased land, including, without limitation, the harvesting of current tree growth, the planting and harvesting of hardwood trees, and the development of wood-based industries;

- Trade and profit from any resulting carbon credits from timber on the leased land;
- Engage in agricultural activities, including the cultivation of biofuel crops (jatropha plant and palm oil trees);
- Explore, develop, mine, produce and/or exploit petroleum, natural gas, and other hydrocarbon resources for both local and export markets, as well as other minerals, and may also engage in power generation activities on the leased land;
- Sublease any portion or all of the leased land or to sublicense any right to undertake activities on the leased land to third parties.[36]

In a situation like this, any belief that a mere proclamation that communities now own their resources would resolve the problem of destructive extraction in Africa amounts to no more than a dream. Getting to that point would require careful and long-drawn knowledge-sharing meetings and consultations. It would require grassroots mobilisations and solidarity.

The web of resistance building across the continent suggests that the strongest thread will be deliberate struggle for democratic accountability. This will require making the community unit the real base of power, beyond the current decentralised delineation of political wards as resource-poor victims of the neoliberal state. In that way, although the immediate struggle may be for a clean environment, for climate justice or for food sovereignty, the ultimate aim is the transformation of society into one in which the people regain true sovereignty.

It would also require political mobilisation to ensure that credible candidates participate in credible elections. Heads of governments staying fastened to their thrones for life will not allow for true reform and change in Africa. Muammar Gaddafi remained the leader of Libya from 1970 until citizens fighting under the air cover of NATO forces chased him off his sumptuous throne in 2011. Meanwhile, Mugabe has plodded on since 1980 and, after taking World Bank advice during the early 1990s and de-industrialising his country's economy, he then drove Zimbabwe to the brink of catastrophe. Africa needs leaders who hear the people and who are elected by the people and who understand the aspirations of the people.

Africa has been exploited almost to death. While her children hack at each other's limbs, others adorn their fingers with the diamonds snatched from their lands. While the blood of her children flows in her numerous oilfields, these fuels roar in engines of the fast life and egregious consumption far off her shores. Mountains and oceans of waste take over Africa while her exploiters dance on piles of dollars and euros. She gasps for breath while transnational rippers and local rapists plunder her lands.

One could be tempted to think that there is nothing left to be rescued in such a despoiled continent. One could throw up one's hands in despair and say well, what a lost continent. Yes, there is a lot that is ugly in the land, but there is still much undeniable beauty, and this must be rescued, nurtured and preserved. Africa deserves a break, a Sabbath of rest. She deserves a space to catch her breath. It is only a united, aware and connected people who will arise and halt the tom toms of death, throw off the parasites and reclaim the land. To do any less may mean getting cooked yet again as the world moves on.

The opportunity for Africa to catch its breath and recover from centuries of despoliation will not happen without a fight. The direction for such a struggle is already seen in the emergence, among others, of networks such as Climate Justice Now! (CJN). We see through CJN that this path will not be elitist, top-down, but broad-based and inclusive; not professional NGO-driven, but seeing itself as a wider socio-environmental movement that does not exclude actors from official circles, especially when such players are not stuck on the creed of officialdom. According to Bond and Dorsey, those environmental specialists espousing neoliberal jargon, 'carry out climate or development advocacy mainly within multilateral institutions or from international NGOs, especially in New York, Washington, London and Geneva, [and] commitments to top-down approaches are held with an almost religious fervor.'[37]

The clear erosion of multilateralism in global negotiations make such spaces arid arenas for any hope of securing actions that would free Africa from the exploitative and hegemonic powers that have seen her as nothing more than a backwater for resources, human and material. Most of Africa's and other poor countries' voting shares within the World Bank are not significant and are stagnating or falling. Even the UN Millennium

Development Goals launched in 2000 and much loved by African governments have proved illusory for Africa partly because the institutions that have brought misery to the continent – the World Trade Organisation (WTO) and the Bretton Woods institutions – play vital intermediary roles for MDG delivery.

The space for empowerment and recovery is clearly in the gathering of the peoples. A potent initial signal for this was laid in the 2010 People's summit on Climate Change held in Cochabamba, Bolivia, under the auspices of President Evo Morales, who showed the world that governments could stand with the people, not against them, and insist on right frameworks for action to be taken.

The 2011 uprisings in North Africa may have somewhat blocked out of view the many stirrings in the rest of Africa below the Sahara desert, but these show important signs that business as usual has a terminal point and that change will come.[38] The underpinnings of the uprisings include rising food prices and this cannot be divorced from the impacts of land grabs by speculators, destructive extraction and appropriation of the people's heritage, climate change, and harsh economic realities all driven by anti-people political actors.

From Dakar to Mogadishu and from Cape Town to Cairo, the peoples of this continent are slowly but surely recovering their voices. Do not mistake the stumping, singing and jumping for a dance party, these are the generators that power the dynamos and set the path for resistance and for change.

We thought it was oil ... but it was blood[39]

The other day
We danced in the street
Joy in our hearts
We thought we were free
Three young folks fell to our right
Countless more fell to our left
Looking up,
Far from crowd
We beheld
Red-hot guns

We thought it was oil
But it was blood

We thought it was oil
But it was blood

Heart jumping
Into our mouths
Floating on
Emotion's dry wells
We leapt in fury
Knowing it wasn't funny
Then we beheld
Bright red pools

We thought it was oil
But it was blood

We thought it was oil
But it was blood

First it was the Ogonis
Today it is the Ijaws
Who will be slain this next day?
We see open mouths
But hear no screams
Tears don't flow
When you are scarred
We stand in pools
Up to our knees

We thought it was oil
But it was blood

We thought it was oil
But it was blood

Dried tear bags
Polluted streams
Things are real
When found in dreams
We see their shells
Behind military shields
Evil, horrible, gallows called oilrigs
Drilling our souls

We thought it was oil
But it was blood

We thought it was oil
But it was blood

The heavens are open
Above our heads

Toasted dreams in
In a scrambled sky
A million black holes
In a burnt out sky
Their pipes may burst
But our dreams won't burst

We thought it was oil
But it was blood

We thought it was oil
But it was blood

They may kill all
But the blood will speak
They may gain all
But the soil will RISE
We may die
And yet stay alive
Placed on the slab
Slaughtered by the day
We are the living
Long sacrificed

We thought it was oil
But it was blood

We thought it was oil
But it was blood

Notes

Chapter 1 Introduction – the pull of Africa

1   Excerpt from the poem 'We thought it was oil but it was blood', from Nnimmo Bassey's collection *We Thought It Was Oil But It Was Blood* (2008, 2nd edition) Ibadan, Kraft Books

2   Pakenham, Thomas (1991) *The Scramble for Africa*, London, Abacus, p. xxiv

3   Marx, Karl (1972) 'The future of British Rule in India', in Karl Marx and Federick Engels, *On Colonialism*, International Publishers, p. 82

4   Tutu, Desmond (2005) *God has a Dream – A Vision of Hope for Our Time*, London, Rider, p. 12

5   Owugah, Lemuel, (2007) *The Politics of Nigeria's Relations with the European Economic Community: From Lagos to Lome*, Port Harcourt, Kemuela Publications, pp. 6–7

6   *Time* magazine (1957) 'United Nations: foursquare for France', 18 February, http://www.time.com/time/magazine/article/0,9171,809106,00.html#ixzz1UqvcGnux

7. Pakenham (1991) p. 30

8   Coleman, James S. (1986) *Nigeria: Background to Nationalism*, Benin City, Broburg & Wistrom, p. 56

9   O'Sullivan, Michael E. (2007) 'Hitler's victims: the German army massacres of black French soldiers in 1940', *Canadian Journal of History*, http://www.britannica.com/bps/additionalcontent/18/27866949/Hitlers-African-Victims-The-German-Army-Massacres-of-Black-French-Soldiers-in-1940#

10 Alie, Joe A. D. (1991) *A New History of Sierra Leone*, Oxford, Macmillan Education, p. 168

11 Meredith, Martin (2006) *The State of Africa – A History of Fifty Years of Independence*, London, Free Press, p. 7

12 Tutu (2005) pp. 25–6

13 Eduardo, Galeano (1997) *Open Veins of Latin America – Five Centuries of the Pillage of a Continent*, New York, NY, Monthly Review Press, p. 80

14 Karl, Marx (1976) *Capital*, Volume 1, New York, NY, Vintage, p. 896

15 Foster, J.B. and Clark, B. (2003) 'Ecological imperialism: the curse of capitalism in the new imperial challenge', in Leo Panitch and Colin Leys (eds) *Socialist Register 2004*, London, Merlin Press, pp.188–9

16 Turner, Terisa E. and Oshare, M.O. (1994) 'Women's uprisings against the Nigerian oil industry in the 1980s' in Terisa E. Turner (ed) *Arise Ye Mighty People – Gender, Class & Race in Popular Struggles*, Trenton, NJ, Africa World Press, pp. 131–2

17 Bond, Patrick (2006) *Looting Africa – the Economics of Exploitation*, Pietermaritzburg, Durban and London, University of KwaZulu-Natal Press and Zed Books, pp. 55–91

18 de Senarclens, Pierre (2007) 'Decolonisation and development, at the origins of the UN Africa geopolitics – Africa in the United Nations system', Paris, no. 25, January–March, p.196
19 Ibid

Chapter 2 Africa is rich

1 Excerpt from the poem 'Yasuni' from Nnimmo Bassey's collection *I Will Not Dance to Your Beat* (2011) Ibadan, Kraft Books
2 Rodney, Walter (1981) *How Europe Underdeveloped Africa* (revised edition), Washington DC, Howard University Press
3 Bassey, Nnimmo (2007) Towards a Political Framework for Food Security and Sustainable Agriculture in Africa – A contribution at the conference on Can Africa Feed Itself? Poverty, Agriculture and Environment – Challenges for Africa, 6–8 June, Oslo, Norway
4 Ernesto 'Che' Guevara (2001) *The African Dream – The Diaries of the Revolutionary War in the Congo*, London, Harvill Press, p. 57
5 Swagler, Matt (2008) 'Behind the war in Congo', *Socialist Worker*, http://socialistworker.org/2008/11/25/behind-war-in-the-congo
6 As quoted in Swagler (2008)
7 'Corpse Award 2005 press coverage', quoted in *Sunday Times*, http://ccs.ukzn.ac.za/default.asp?2,40,5,1168
8 Swagler (2008)
9 Block, Robert (1997) 'U.S. firms seek deals in Central Africa,' *Wall Street Journal*, p. A17, 14 October
10 Montague, D. and Berrigan, F. (2001) 'The business of war in the Democratic Republic of Congo: who benefits?', July–August, World Policy Institute
11 Ibid
12 McCartney, L. (1998) *Friends in High Places: The Bechtel Story: The Most Secret Corporation and how it Engineered the World*, Simon & Schuster
13 Bangura, Ahmed Ojulla (2008) 'Environmental degradation in Kono area', *Concord Times* (Freetown), 21 April, http://allafrica.com/stories/200804212037.html
14 *Sierra Express Media* (2009) 'APC allows town mining in Kono', 5 October, http://www.sierraexpressmedia.com/archives/527
15 The company profile is available at http://www.koiduholdings.com/about_company_profile.html
16 Network Movement for Justice and Development (2004) 'The Koidu Kimberlite Project – Is Koidu Holdings Ltd above the law?', press statement, 11 February, http://www.minesandcommunities.org/article.php?a=694
17 Bank Information Centre (2004) 'Koidu Holdings Limited denies any wrongdoing or violation of EIA policy', 14 January, http://www.bicusa.org/EN/Article.705.aspx
18 Koidu Holdings (2011) 'Key facts', http://www.koiduholdings.com/company-key-facts.php

19 Alie, Joe A. D. (1990)

20 Akabzaa, T.M., Seyire, J.S. and Afriyie K. (2007) 'The glittering façade – effects of mining activities in Obuasi and its surrounding communities', Accra, Third World Network-Africa (TWN-Africa), p. 1; see also Kairos (2004) *Africa's Blessing, Africa's Curse: The Legacy of Resource Extraction in Africa*, Accra, TWN-Africa

21 *The Economist* (2009) *Pocket World Figures 2008*, London, *The Economist*, pp. 52–3

22 *Oil Review Africa* (2009) 'Nexans wins $9.8 million topside contract for Usan FPSO offshore Nigeria', issue 3, p. 30

23 *Oil Review Africa* (2009) 'World's first FDPSO now in operation', issue 3, p. 27

24 Ghazvinian, John (2009) *Untapped – The Scramble for Africa's Oil*, Houghton Mifflin Harcourt, pp. 9–11

25 Ibid, p. 214

26 Woods, Emira (2009) 'Obama visits Africa's "Oil Gulf"', Inter-Press Service, 12 July, http://www.ips-dc.org/articles/ obama_visits_africas_oil_gulf

27 'Resource curse' (2010) *London Review of Books*, 8 July, http://www.lrb. co.uk/blog/2010/07/08/khadija-sharife/5299/

28 Akabzaa et al (2007) p. 7

29 Ibid

30 Ibid

31 Kairos (2004)

32 Akabzaa et al (2007) p. 14

33 AngloGold could make these decisions because virtually the entire Obuasi is their concession area. They can mine where they please and move whatever stands in their way.

34 Norwegian University of Life Sciences (2009) 'Tanzania: trace metal concentrations in soil, sediments, and waters', http://www.africafiles.org/ article.asp?ID=22228

35 Ibid

36 Reuters (2009) 'Zambia opposition moves to block Chinese mine deal', 24 June, http://www.miningweekly.com/article/zambia-president-says- willing-to-review-mine-taxes-2009-06-24

37 Shacinda, Shapi (2009) 'Zambia social programmes may suffer as copper mine revenues fall', Reuters, 18 June, http://www.mineweb.com/ mineweb/view/mineweb/en/page504?oid=85133&sn=Detail

38 EITI (2010) 'Citizens in 24 countries are now able to see the revenues from resources', November, http://eiti.org/news-events/billions-revenues- are-being-reported.

39 Sharife, Khadija (2011) 'Transparency hides Zambia's billions', Al Jazeera June 18, http://english.aljazeera.net/indepth/opinion/2011/06/ 20116188244589715.html

40 EITI (2011) 'Zambia becomes 26th country to publish an EITI report', 23 February, http://eiti.org/news-events/zambia-becomes-26th-country- publish-eiti-report

41 Financial Action Task Force (2011) 'Tackle illicit financial flows for the new bottom billion', 13 July, http://www.financialtaskforce.org/2011/07/13/tackle-illicit-financial-flows-for-the-new-bottom-billion/

42 Sharife, Khadija (2011)

43 *Christian Science Monitor* (2011) 'Zambia: new President Sata sets new mining rules for China', 28 September, http://www.csmonitor.com/World/Africa/2011/0928/Zambia-s-new-President-Sata-sets-new-mining-rules-for-China.

44 Financial Secrecy Index (2010) 'The British connection', http://www.financialsecrecyindex.com/documents/FSI%20-%20The%20British%20Connection.pdf

45 *The Africa Report* (2010) 'Building Africa's tax havens', 2 December, http://www.theafricareport.com/typerighter/index.php?post/2010/12/02/Building-African-tax-havens

46 China Offshore (n.d.) 'The Gateway: Mauritius is at the center of indian ocean trade flows connecting India, Africa, and China', http://www.chinaoffshore.com.hk/the-gateway.html

47 Financial Secrecy Index (2010) 'Mauritius', http://www.secrecyjurisdictions.com/PDF/Mauritius.pdf

48 Letter from Rio Tinto plc to the US Securities and Exchange Commission, http://www.sec.gov/comments/s7-42-10/s74210-44.pdf

49 Palitza, Kristin (2009)'Economies must diversify, reduce focus on mining', 10 June, http://www.ipsnews.net/africa/

50 BBC (2009) 'Guinea confirms huge China deal', 13 October

51 The National (2008) 'Group blasts company for mining deal with Eritrea', http://www.thenational.ae/article/20081226/FOREIGN/883029914

52 Kairos (2004) p. 28

53 The Citizen Reporter (2009) 'Experts predict tripling of income from mining', 8 June, http://thecitizen.co.tz/newe.php?id=12973

54 Statement issued in Accra, Ghana (2009) 13 August

55 Festus Iyayi, in comments on what civil society must do to raise awareness on the political underpinnings of the environmental degradation

56 *The Guardian* (2009) 'Great expectations in Uganda over oil discovery', 2 December, http://www.guardian.co.uk/katine/2009/dec/02/oil-benefits-rural-uganda

57 Ibid

58 Belfer Center (2011) 'Russia in Review', 8 April, http://belfercenter.ksg.harvard.edu/publication/20925/russia_in_review.html

59 Interview with Khadija Sharife (August 2011)

Chapter 3 The wheels of progress

1 Excerpt from the poem 'Walking blind' from the Nnimmo Bassey's collection, *I Will Not Dance to Your Beat* (2011) Ibadan, Kraft Books

2 Pakenham, Thomas (1991) *The Scramble for Africa*, London, Abacus, p. xxv

3   Jarecki, Eugene (2006) *Why We Fight*, Sony Classic Pictures and Charlotte Street Film

4   Chatterjee (2009) *Halliburton's Army – How a Well-Connected Texas Oil Company Revolutionized the Way America Makes War*, New York, Nation Books, p. 213

5   Landau, Saul (2007) *A Bush and Botox World*, California, Counterpunch, p. 60

6   Ibid, p. 61

7   United Nations Conference on Trade and development (UNCTAD) (2008) *Economic Development in Africa 2008. Export Performance Following Trade Liberalization: Some Patterns and Policy Perspectives*, New York and Geneva, United Nations, http://www.unctad.org/en/docs/aldcafrica2008_en.pdf

8   Moyo, Dambisa (2009) *Dead Aid – Why Aid Is Not Working and How There Is Another Way for Africa*, London, Allen Lane

9   Calderisi, Robert (2007) *The Trouble With Africa: Why Foreign Aid Isn't Working*, New Haven and London, Yale University Press

10 Moyo (2009) p. 15

11 Ibid, pp. 15–16

12 Bond, Patrick (2006) *Looting Africa – The Economics of Exploitation*, Pietermaritzburg, University of Kwa Zulu-Natal Press, pp. 33–4

13 Fontanel, Jacques and Biays, Joël-Pascal (2007) 'Africa and the IMF', *Africa Geopolitics*, January–March, p. 227

14 Calderisi (2007) pp. 208–17

15 Bond, Patrick (1999) 'Globalization, pharmaceutical pricing and South African health policy: managing confrontation with US firms and politicians', *International Journal of Health Services*, vol. 29, no. 4

16 Bond (2006) p. 35

17 Ibid

18 Cited in SA Institute for International Affairs e-Africa (2004) May

19 Chomsky, Noam (2007) *Failed States – The Abuse of Power and the Assault on Democracy*, London, Penguin Books, p. 25

Chapter 4 The steps of the advisers

1   Excerpt from the poem 'Laguna Guatavita', from Nnimmo Bassey's collection *We Thought It Was Oil But It Was Blood* (2008, 2nd edition) Ibadan, Kraft Books

2   Klein, Naomi (2009) 'Why we should banish Larry Summers from public life', 19 April, http://www.naomiklein.org/articles/2009/04/why-we-should-banish-larry-summers-public-life

3   Summers, Lawrence (1991) 'Let them eat pollution', 12 December, quoted by Aaron Petcoff, incoming economic adviser, http://aaronpetcoff.com/2009/01/02/let-them-eat-pollution/. This leaked memo was first published by *The Economist*, 8 February 1992

4   Guest, Robert (2004) *The Shackled Continent: Africa's Past, Present and Future*, London, Macmillan, p. 20

5   Bolton, Giles (2008) *Aid and Other Dirty Business – an Insider Reveals How Globalisation and Good Intentions Have Failed the World's Poor*, London, Ebury Press, p. 218

6   'African Alternative Framework to Structural Adjustment Programs for Socio-Economic Recovery and Transformation' (1989) http://www.africaaction.org/african-initiatives/aaf3.htm

7   Chomsky, Noam (2003) *Radical Priorities*, Otero, C.P. (ed), Oakland, CA, AK Press, p. 279

8   Ibid

9   Stiglitz, Joseph (2000) 'What I learned at the world economic crisis', *The Insider, The New Republic*, 17 April, cited in Shah, Anup (2008) 'Structural adjustment—a major cause of poverty', *Global Issues*, updated 29 October, http://www.globalissues.org/article/3/structural-adjustment-a-major-cause-of-poverty

10   Bretton Woods Project Update (2001) 'PRSPs just PR, say civil society groups', http://www.brettonwoodsproject.org/art-15999

11   The World Bank (n.d.) http://web.worldbank.org/WBSITE/EXTERNAL/NEWS/0,,contentMDK:20040942~menuPK:34480~pagePK:34370~theSitePK:4607,00.html

12   World Bank (2009) 'Debt relief', http://web.worldbank.org/WBSITE/EXTERNAL/EXTSITETOOLS/0,,print:Y~isCURL:Y~contentMDK:20147607~menuPK:344191~pagePK:98400~piPK:98424~theSitePK:95474,00.html

13   Millet, Damien and Toussaint, Eric (2004) *Who Owes Who? – 50 Questions About World Debt*, London, Zed Books, p. 90

14   See http://go.worldbank.org/4IMVXTQ090

15   Bloomberg (2011) 'IMF electoral math doesn't add up', 25 May, http://www.bloomberg.com/news/2011-05-25/the-imf-s-electoral-math-doesn-t-add-up.html

16   Carnegie Council (2002) 'The mystery of capital', 8 May, http://www.carnegiecouncil.org/resources/publications/morgenthau/99.html

17   *Global Policy Journal* (2011) 'Profile: Hernando de Soto', http://www.globalpolicyjournal.com/practitioners-advisery-board/hernando-de-soto

18   Vicky, Alain (2011) 'Who owns Buganda', August, *Le Monde Diplomatique*

19   *World Policy Journal* (2011), 'This land is your land', summer, http://www.worldpolicy.org/journal/summer2011/this-land-is-your-land

20   Vicky, Alain (2011)

21   de Soto, Hernando (2000) *The Mystery of Capital*, New York, NY, Basic Books

22   Guest (2004) pp. 58–61

23   Tran, Mark (2009) 'UN denies complicity in Congo war crimes', 11 November, *The Guardian*, http://www.guardian.co.uk/world/2009/nov/11/congo-un-rebels-rwanda-kimia

24   Pan African News Wire (2009) 'Clinton threatens Eritrea over US failure in Somalia', 9 August, http://panafricannews.blogspot.com/2009/08/clinton-threatens-eritrea-over-us.html

25   Friends of the Earth (2009) 'ArcelorMittal in Liberia – problems to iron out', in *ArcelorMittal; Going Nowhere Slowly – A Review of the Global Steel Giant's*

Environmental and Social Impacts in 2008-2009, May, http://www.foeeurope.
org/corporates/Extractives/arcelormittal_going_nowhere_web.pdf

26 Murphy, Richard (2007) 'Mittal Steel did the right thing: will Firestone?'
    1 May

27 Global Witness (2007) 'Heavy Mittal', 2 October, http://www.
    globalwitness.org/library/heavy-mittal

28 Friends of the Earth (2009)

29 Williams, Stephen (2009) 'A frontier too far?', *Oil Review Africa*, no. 3

30 Urquhart, Sam (2009) 'Fragmenting Sudan the Jarch way',
    Hidden Paw, 3 March, http://szamko.wordpress.com/2009/03/06/
    fragmenting-sudan-the-jarch-way/

31 Jarch (2011) 'Company overview' http://www.jarchcapital.com/company-
    overview.php

32 Ibid

Chapter 5 Destructive extraction

1 Excerpt from the poem 'Gas flares' by Nnimmo Bassey

2 Human Rights Watch (1999) *The Price of Oil – Corporate Responsibility and
   Human Rights Violations in Nigeria's Oil Producing Communities*, New York,
   NY, Human Rights Watch, p. 202

3 Okonta, Ike and Douglas, Oronto (2001) *Where Vultures Feast – 40 years
   of Shell in the Niger Delta*, Benin City, Environmental Rights Action and
   Friends of the Earth Nigeria, p. 142

4 Steiner, Richard (2008) *Double Standards? International Best Practice
   Standards to Prevent and Control Pipeline Oil Spills, Compared with Shell
   Practices in Nigeria*, Amsterdam, Milieudefensie

5 Sheehan, Paul (2001) 'Corporate spin and lies: a spymaster's lament, and
   a warning to us all', *Sydney Morning Herald*, http://www.commondreams.
   org/views01/0228-02.htm

6 Lewis (1996) 'Blood and oil: a special report: after Nigeria represses, Shell
   defends its record', *The New York Times*, cited in Rowell, Andrew (1996)
   Green Backlash – Global Subversion of the Environment Movement, London,
   Routledge, p. 293

7 Christian Aid (2004) *Behind the Mask – The Real Face of Corporate Social
   Responsibility*, p. 30

8 Pigging Products and Services Association (PPSA) (2004) 'Using
   benchmarking to optimise the cost of pipeline integrity management',
   cited by Richard Steiner (2008) *Double standards?: International Best
   Practice Standards to Prevent and Control Pipeline Oil Spills, Compared with
   Shell Practices in Nigeria*, Amsterdam, Friends of the Earth Netherlands

9 Cellarius, Richard (2009) 'The environmental implications of
   unconventional oil production and exploration (Artic and Canada Oil
   [tar] sands)', 13 October, presented at the conference on Extractive
   Industries: Blessing or Curse? Social Impacts of Oil and Gas Industry,
   Brussels, Belgium

10 Okorodudu-Fubara, Margaret T (1998) *Law of Environmental Protection –
    Materials and Texts*, Ibadan Caltop Publications, p. 815

11 Okonto and Douglas (2001)

12 NOSDRA was established by act of the National Assembly on 18
    October 2006 to provide an institutional framework to coordinate and
    implement the National Oil Spill Contingency Plan (NOSCP) for Nigeria
    in accordance with the International Convention on Oil Pollution
    Preparedness, Response and Cooperation (OPRC) 1990, to which Nigeria
    is a signatory. The agency is mandated to ensure timely, effective and
    appropriate response to all oil spills as well as ensuring clean up and
    remediation of oil impacted sites.

13 Nigerian News Service reporting from *Daily Independent* (2009)
    'Nigeria records 2,122 oil spills in four years', 6 October, http://www.
    nigeriannewsservice.com/index.php/Your-Naira/Nigeria-Records-2122-
    Oil-Spills-In-Four-Years.html

14 Based on data from the Nigerian Department of Petroleum Resources
    (DPR)

15 United Nations Environment Programme (UNEP) (2011) *Ogoniland Oil
    Assessment Reveals Extent of Environmental Contamination and Threats to
    Human Health*, http://www.unep.org/newscentre/
    Default.aspx?DocumentID=2649&ArticleID=8827&l=en

16 Quoted in Bassey, Nnimmo (2011) 'The agony of Ogoni', *NEXT*
    newspaper, Lagos, 11 August, http://234next.com/csp/cms/sites/Next/
    Money/5738479-146/story.csp

17 Moody, Roger (1998) *Out of Africa: Mining in the Congo Basin*, IUCN The
    Congo Basin – Human and Natural Resources, Amsterdam, p. 137

18 Reguly, Eric (2008) 'Rio Tinto tries to sidestep elephant in the room',
    14 July, http://www.theglobeandmail.com/servlet/story/LAC.20080714.
    RREGULY14/TPStory?cid=al_gam_globeedge

19 *Times* (2010) 'Fabulous wealth in Marange diamonds', 8 August,
    http://www.timeslive.co.za/africa/article591181.ece/
    Fabulous-wealth-in-Marange-diamonds

20 BBC (2011) 'Marange diamond field torture camp discovered', *Panorama*,
    8 August, http://www.bbc.co.uk/news/world-africa-14377215

21 See *Africa–Asia Confidential: Zimbabwe Country Profile*

22 Levkowitz, L., Ross, M. and Warner, J. (2009) 'The 88 Queensway Group:
    a case study in Chinese investors' operations in Angola and beyond',
    prepared for US–China Economic & Security Review Commission.

23 Ibid

24 Swain, J. (2011) 'ZANU in shadow of elusive magnate', 12 March,
    http://www.businesslive.co.za/africa/2011/03/12/
    zanu-in-shadow-of-elusive-magnate

25 Brautigam, D. (2009) *Dragon's Gift: The Real Story of China in Africa*,
    Oxford, Oxford University Press

26 *Africa–Asia Confidential* (2009) 'China Sonangol targets Harare's gold and
    oil', November

27 Reuters (2011) 'China's Anjin Zimbabwe diamond output hits 1 million

carats', 7 April, http://www.businesslive.co.za/Feeds/reuters/2011/04/07/china-s-anjin-zimbabwe-diamond-output-hits-1-million-carats

28 Interview with Khadija Sharife (June 2011)

29 Zimbabwe Metro (2010) 'Arrested diamond researcher "set up" by KP monitor', 3 June, http://www.zimbabwemetro.com/headline/arrested-zimbabwean-diamond-researcher-%E2%80%98set-up%E2%80%99-by-kp-monitor/

30 *The Africa Report* (2011) 'Companies profit from toxic water', March, http://www.theafricareport.com/component/content/article/54/5138186.html

31 Ibid

32 Ibid

33 Africa Earth Observatory Network (AEON) (2010) 'H_2O-CO_2 energy equations for SA', 22 November, http://www.aeon.uct.ac.za/news/news.php?newsId=29&start=6

34 Wikileaks would later disclose that, according to a US embassy official, 'A contact at the International Finance Corporation in Washington DC said that someone had paid $US7 million to someone in the presidency as a bribe to get the Rio Tinto contract cancelled.' The rumour may or may not have been true.

35 BIC (2008) 'IFC considers record mining investment in Guinea as its Simandou concession comes under dispute' 11 August at http://www.bicusa.org/en/Article.3871.aspx

36 Hotter, A. and Matthews, R.G. (2009) 'Rio Tinto, Guinea spar over mining rights – miner told to remove equipment, cede half of mineral-rich area', 24 July, http://online.wsj.com/article/SB124838931608377349.html

37 Rio Tinto (2009) 'Rio Tinto Simandou project information', July, www.riotintosimandou.com

38 *Financial Times* (2011) 'Guinea to review mining licences', 4 March, http://www.ft.com/intl/cms/s/0/5ae818ec-469e-11e0-967a-00144feab49a.html#axzz1XDNQGJuq

39 *Africa Confidential* (2011) 'Conde drives a hard bargain', 8 July

40 Souare, I. (2009) 'Explaining the December 2008 military coup d'etat in Guinea', *Conflict Trends*, no. 1

41 *Africa Confidential* (2011) 'Conde drives a hard bargain', 8 July

42 *Africa Confidential* (2010) 'Votes and the mining house', 28 May

43 Al Jazeera (2011), 'Guinea's president residence attacked', 19 July, http://english.aljazeera.net/news/africa/2011/07/201171974410871681.html

44 Africa News (2011) 'Guinea: security forces repress demonstration', 28 September, http://www.africanews.com/site/Guinea_Security_forces_repress_demonstration/list_messages/39892

45 Banro Corporation (2009) 'Banro Foundation and the Luhwindja community celebrate handover of new high school and potable water system', press release, Toronto, 11 May

46 John Lasker (2009) 'Digging for gold, mining corruption', 29 October, http://canadiandimension.com/articles/2565/>http://canadiandimension.com/articles/2565/

47 *The Guardian* (2009) 'Germany arrests Hutu extremist group leaders', Lagos, 18 November, p. 10

48 Almeida, Henrique (2009) 'Extra troops as illegal diamond miners flood over DRC/Angola border', Reuters, 19 May

49 The Congo Conflict Minerals Act sponsored by Richard Durbin, Senate Majority Whip, along with Senators Sam Brownback, Republican from Kansas, Russell Feingold, Democrat of Wisconsin, and Charles Schumer, Democrat of New York

50 Knight, Danielle (2009) 'DR-Congo: U.S. Congress moving to track "conflict minerals"', Inter-Press Service, 15 May, http://www.ipsnews.net/africa/nota.asp?idnews=46868

51 Lasker, John (2009)

52 Reuters (2009) 'DRC, First Quantum must relaunch talks – minister', 9 October, http://www.miningweekly.com/article/drc-first-quantum-must-relaunch-talks---minister-2009-10-09

53 IRIN (2006) 'Zambia: Kabwe, Africa's most toxic city', 9 November, http://www.irinnews.org/Report.aspx?ReportId=61521

54 IRIN (2008) 'Paying the price for mining', 15 February, http://www.irinnews.org/printreport.aspx?reportid=76780

55 groundWork (2008) 'Wasting the Nation – making trash of people and places', Pietermaritzburg, groundWork Report

Chapter 6 Climate chaos and false solutions

1 'If climate change were little change' is from Nnimmo Bassey's collection *I Will Not Dance to Your Beat* (2011) Ibadan, Kraft Books

2 Greenpeace (2005) 'Climate change: a burden Africa cannot afford', 6 July, http://www.greenpeace.org.uk/blog/climate/climate-change-a-burden-africa-cannot-afford

3 Intergovernmental Panel on Climate Change (IPCC) (2007) 'Summary for policymakers', in *Climate Change 2007: Impacts, Adaptation and Vulnerability. Contribution of working Group II to the Fourth Assessment Report of the Intergovernmental Panel on Climate Change*, Cambridge, Cambridge University Press.

4 Calderisi, Robert (2007) *The Trouble With Africa – Why Foreign Aid Isn't Working*, New Haven, Yale University Press, pp. 126–7

5 Collier, Paul et al (2008) 'Climate change and Africa', 6 May, http://users.ox.ac.uk/~econpco/.../pdfs/ClimateChangeandAfrica.pdf

6 Godrej, Dinyar (2006) 'The no-nonsense guide to climate change', *New Internationalist*, p. 14

7 Lohmann, Larry (2009) 'Neoliberalism and the calculable world: the rice of carbon trading', in Birch, K., Mykhnenko, V. and Trebeck, K. (eds) (forthcoming) *The Rise and Fall of Neoliberalism: The Collapse of an Economic Order?* London, Zed Books. It should be noted Al Gore eventually became a carbon market player himself.

8 Monbiot, George (2009) 'A self-fulfilling prophecy', *The Guardian*, 17 March

9   In a *Newsweek* interview, 'Countdown to Copenhagen' (November 16, 2009), for instance, Kevin Rudd, prime minister of Australia claimed: 'There are a number of well-established technologies in the CCS field. What is missing, however, is sufficient projects at scale, by which I mean electricity projects generating 500 megawatts. The G8 has committed to having 20 of these projects online by 2020, but there are none of these as of 2009.'

10 Carbon Trade Watch (2007) *The Carbon Neutral Myth – Offset Indulgences for your Climate Sins*, Amsterdam, Transnational Institute, p. 54

11 Gilbertson, Tamra and Reyes, Oscar (2009) *Carbon Trading – How it Works and Why it Fails*, Uppsala, Dag Hammarskjold Foundation, p. 10

12 Bond, Patrick (2010) 'Maintaining momentum after Copenhagen's Collapse: seal the deal or "Seattle" the deal?', *Capitalism Nature Socialism*, vol. 21, no. 1, pp. 14–27

13 This is the position of Friends of the Earth International and other climate-justice campaigning civil society groups.

14 According to Wikipedia, the People's Republic of China is the largest consumer of coal in the world, and is about to become the largest user of coal-derived electricity, getting 1.95 trillion kilowatt-hours per year, or 68.7 per cent of its electricity from coal as of 2006 (compared to 1.99 trillion kilowatt-hours per year, or 49 per cent for the US). Hydroelectric power supplied another 20.7 per cent of China's electricity needs in 2006. With approximately 13 per cent of the world's proven reserves, China has enough coal to sustain its economic growth for a century or more even though demand is currently outpacing production. China's coal mining industry is the deadliest in the world and has the world's worst safety record where an average of 13 people die every day in the coal pits, compared to 30 per year for coal power in the United States. See http://en.wikipedia.org/wiki/Coal_power_in_China

15 Bassey, Nnimmo (2008) 'Interrogating official mechanisms for tackling climate change', June at www.eraction.org

16 International Institute for Applied Systems Analysis (2008) 'Food security and sustainable agriculture – the challenges of climate change in sub-Saharan Africa', 8 May at a side event of the United Nations Commission on Sustainable Development-16 at the UN, New York

17 Anderson, Teresa (2009) 'Email on the ABN lists', 11 November

18 Lee, Josie (2009) 'Differences over indigenous peoples' rights and forest conversion in REDD-plus', 9 October, TWN Bangkok News Update, no.18, http://www.twn.org.sg

Chapter 7 Leaving the Niger Delta's oil in the soil

1   Excerpt from the poem 'Justice now' from Nnimmo Bassey's collection *I Will Not Dance to Your Beat* (2011) Ibadan, Kraft Books

2   Ojo, Eric (2009) 'Bankole laments illegal oil bunkering in Niger Delta – challenges security agencies on leakages in public funds', *Business Day,* 12 November

3   Amnesty Press Statement (2011) Henry Ugbolue (head of media and communication) 6 September

4   See http://www.iht.com/articles/2008/07/31/business/oil.php

5   World Socialist Website (2008) 'Oil giants report massive profits' 6 August, http://wsws.org/articles/2008/aug2008/oil-a06.shtml. This report indicated that 'The major US oil companies appear headed for a combined $160 billion in profits for 2008. That compares to $123 billion in 2007. Exxon and other oil companies have rewarded their CEOs with multi-billion dollar payouts. Last year Exxon CEO Rex Tillerson cashed in $16.1 million in stock options in addition to his $1.75 million salary. He also received a $3.36 million bonus. Conoco Chairman James Mulva received $31.3 million last year.'

6   Quoted in ERA/CJP (2005) *Gas Flaring in Nigeria: A Human Rights, Environmental and Economic Monstrosity*, Amsterdam, June. This booklet can be found at both www.climatelaw.org and at www.eraction.org

7   Environmental Rights Action (ERA) (2008) 'ERA fact sheet on gas flaring', December

8   *Multinational Monitor* (2007) 'The end of oil' (editorial), January–February, p. 6. This issue of the *Multinational Monitor* illustrates, among others, that the 'Corporate control of energy policy and energy resources, especially in the United States, the country that consumes more energy than any other, is the single greatest obstacle to slow and hopefully reverse the world's headlong rush to disaster.'

9   At a growth rate of 2.025 per cent. See https://www.cia.gov/library/publications/the-world-factbook/print/ni.html

10 Centre for Research on Multinational Corporations (SOMO) (2008) 'Taxation and financing for development', SOMO Paper, October.

11 New reports abound to show this. See for example, Ezeobi, Obinna (2008) 'FG suspends oil bid rounds', *The Punch* , 23 August, http://www.punchontheweb.com/Articl.aspx?theartic=Art200808231593070

Chapter 8 Swimming against the tide, connected by blood

1   'They charged through the mounted troops' is a poem from Nnimmo Bassey's collection *I Will Not Dance to Your Beat* (2011) Ibadan, Kraft Books

2   Quoted in Meredith, Martin (2006) *The State of Africa – A History of Fifty Years of Independence*, London, Free Press. Meredith reaches the conclusion that it was sentiments such as this that proved ruinous for Mozambique and eventually led to a civil war. We reach a different conclusion about this speech. It was a frank declaration and pointed to the optimal way for a sovereign state with a genuinely post-colonial

orientation. Those who claim to be citizens of a nation ought to agree to what would bring about the common good. The history of neo-colonial Africa is replete with fractious reactions to nationalist moves and in this way the continent has been pushed into a cul-de-sac from where the escape route is blocked by huge economic, political and social barriers.

3 Weisman, Alan (2008) *The World Without Us*, London, Virgin Books, pp. 72–3

4 Dahomey is today known as Benin Republic

5 Wikipedia, 'Dahomey Amazons', http://en.wikipedia.org/wiki/Dahomey_Amazons

6 Turner, Teresa and Brownhill, Leigh (2005) 'Why women are at war with Chevron: Nigerian subsistence struggles against the international oil industry', in Turner, Teresa and Brownhill, Leigh *The New Twenty-First Century Land and Oil Wars: African Women Confront Corporate Rule*, New York, NY, International Oil Working Group

7 Environmental Rights Action (ERA) (2002) 'Protesting women continue occupation of Chevron flow-stations', ERA Field Report, 22 July

8 Ogoni Bill of Rights (2nd edition) (1992) Port Harcourt, Saros International. The Bill of Rights was first produced in an unpublished form in 1990

9 The repression that followed Operation Climate Change inspired my poem 'We thought it was oil but it was blood' (from the book of the same title, Kraft Books, 2002), an extract from which appears at the beginning of Chapter 1 of this book

10 Oilwatch International (2006) *Chevron: The Right Hand of the Empire – Urgent Information on ChevronTexaco*, Quito, Oilwatch International

11 ERA (2008) 'Chevron trial: Ilaje will appeal ruling', press statement, Lagos, 3 December

12 United States District Court for the Northern District of California (2009) Larry Bowoto et al., plaintiffs, v. Chevron Corporation et al., defendant, Order Denying Bill of Costs, No. C 99-02506 SI, p. 2

13 Ibid, p. 3

14 See United States District Court for the Northern District of California (2009) for details

15 See http://www.ienearth.org/keystone-xl-pipeline.html?utm_medium=email&utm_source=MyNewsletterBuilder&utm_content=109788589&utm_campaign=Lets+unite+with+one+voice+Keystone+XL+Pipeline+Hearings+1411016627&utm_term=

16 Klein, Naomi (2007) *The Shock Doctrine*, London, Penguin Books

17 Bassey, Nnimmo (2007) *Oil, Environment and Crisis Economics*, http://www.pambazuka.org/en/category/comment/44280

18 ERA (2009) 'ERA and FoE's visit to Gbaramatu displaced persons', ERA Field Report no. 207, 12 June

19 Adebayo, Sola (2009) 'Ex-militant leaders suspend talks with FG', *The Punch* (Lagos), 8 November, p. 7

20 Ibid

21 Owuamanam, Jude (2009) 'Bunkering: Uduaghan blames oil firms, multinationals', *The Punch* (Lagos), 8 November, p. 5

22 Ojo, Eric (2009) 'Bankole laments illegal oil bunkering in Niger Delta – challenges security agencies on leakages in public funds', *Business Day*, 12 November, p. 8

23 Ehirim, Chuks (2009) 'How naval officers aid Bunkering', *National Daily* (Lagos), 16–22 November

24 Saidou, Djibril (2009) 'Niger President Tandja grants Tuareg rebels amnesty', *Sahel Says*, 26 October http://www.bloomberg.com/apps/news?pid=20601116&sid=a.9TKAMA7DYU

25 Maass, Peter (2009) *Crude World: The Violent Twilight of Oil*, London, Allen Lane

26 In 1990 Umuechem community was sacked by mobile policemen and 500 houses were burnt down and 80 persons were killed simply because the people dared to organise a protest against the exploitation of their resources with none of the benefits accruing to them.

27 The Odi massacre occurred in November 1999 under the presidency of Chief Obsanjo; 2,843 citizens including children and the aged, were killed. For details see ERA, *A Blanket of Silence* at www.eraction.org

28 Sams, Ngozi (2009) 'Manage your oil wealth well, Venezuela urges Nigeria', *NEXT* newspaper, 24 November, http://234next.com/csp/cms/sites/Next/Money/Business/5485512-147/story.csp

29 Dembélé, Demba Moussa (2008) 'Sankara 20 years later: A tribute to integrity' Pambazuka News, no. 402, http://pambazuka.org/en/category/features/51193

30 Ibid

31 Ibid

32 Jacobs, Sean Wednesday (2008) 'Sankara: daring to invent Africa's future', *The Guardian* (UK), 15 October, http://www.guardian.co.uk/profile/seanjacobs, and Cheriff, M.S.Y. (2007) 'Thomas Sankara: chronicle of an organized tragedy', Africa News, 17 September, http://www.africanews.com/site/list_messages/11528

33 The Vanguard (2009) '10 Percent Oil Revenue for Niger-Delta Communities' 20 October, at http://allafrica.com/stories/200910200994.html

34 Douglas, Oronto (2009) 'Niger Delta: defining the community in an age of ten percent', *The Nation* (Lagos), 8 November,http://thenationonlineng.net/web2/articles/24636/1/Niger-Delta-Defining-the-community-in-an-age-of-ten-percent/Page1.html

35 *Nigerian Tribune* (2009) 'Niger Delta communities demand interpretation of 10 percent equity for oil host communities', 16 November, http://www.tribune.com.ng/16112009/news/news11.html

36 Oakland Institute (2011) 'Understanding land investment deals in Africa – Nile Trading and Development Inc. in South Sudan', Land Deal Brief, June, http://media.oaklandinstitute.org/sites/oaklandinstitute.org/files/OI_Nile_Brief_0.pdf

37 Bond, Patrick and Dorsey, Michael K. (2010) 'Anatomies of environmental knowledge and resistance: diverse climate justice movements and waning eco-neoliberalism', *Journal of Australian Political Economy*, no. 66
38 See Manji, Firoze and Eline, Sokari (eds) (2012) *African Awakening: The Emerging Revolutions*, Oxford, Pambazuka Press
39 From the collection by Nnimmo Bassey (2002) *We Thought It Was Oil But It Was Blood*, Ibadan, Kraft Books

Index

Earth Grab: Geopiracy, the New Biomassters and Capturing Climate Genes

ETC Group, led by Diana Bronson, Hope Shand and Jim Thomas

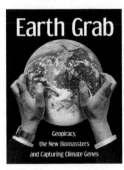

Earth Grab

Geopiracy,
the New Biomassters
and Capturing Climate Genes

2011
paperback
978-0-85749-044-5
also available in pdf format

Despite being at the helm of a system responsible for the climate and energy crisis, global corporations are now taking the lead in generating 'solutions' for the planet's recovery. From three perspectives, *Earth Grab* exposes the dangers of allowing those who are motivated by profit to control this critical process.

'Geopiracy' shows the disastrous impacts that Northern geoengineered 'quick fixes' have on ecosystems and peoples of the global South.

'The New Biomassters' exposes, as corporations seek to switch from oil- to plant-based energy, how a biomass economy threatens and accelerates land grabs in the South.

Lastly, 'Capturing Climate Genes' shows how 'climate-ready crops' will allow agribusiness to expand into the lands peasant farmers currently cultivate, feeding only the gluttony of corporate shareholders for profits.

'These three groundbreaking reports pull back the curtain on disturbing technological and corporate trends that are already reshaping our world and that will become crucial battlegrounds for civil society in the years ahead.'

Dr Vandana Shiva, founder of the Research Foundation for Science, Technology and Ecology

 Order your copy from www.pambazukapress.org

Food Rebellions! Crisis and the Hunger for Justice

Eric Holt-Giménez and Raj Patel

Food Rebellions! is a powerful handbook both examining our vulnerable food systems to reveal the root causes of the food crisis and proposing solutions. Democratising food systems can end hunger and poverty. Sustainable approaches to production must be supported and spread. In local, national and international policy arenas we need dialogue, transparency and a change to the 'rules' currently holding back agroecological alternatives.

2009
paperback
978-1-906387-30-3
also available in pdf, epub
and Kindle formats

'As *Food Rebellions!* demonstrates, ... using traditional farming methods enhances environmental conservation and preserves local biodiversity ... key to the survival of the many African families headed by women.'

Wangari Maathai, Nobel Peace Prize winner

Food Sovereignty

Edited by Hannah Wittman, Annette Aurélie Desmarais
and Nettie Wiebe

2011
paperback
978-0-85749-029-2

With increasing hunger in the world,
especially among marginalised populations
in both the North and South, the current
high-input, industrialised, market-driven food
system is failing.

Advocating a practical, radical change to
the way our food system operates, the authors
argue that putting control in the hands of
those who produce the world's food rather
than corporate executives is the means to
providing for the food needs of all people.

Contributors, including Raj Patel, Walden
Bello, Philip McMichael, Miguel Altieri and
Eric Holt-Giménez, show in both conceptual
and case study terms that food sovereignty
results in increased production, safe food that
reaches those who are in the most need and
agricultural practices that respect the earth.

'For anyone wishing to look behind the usual strategies
advanced for increasing food production and to consider radical
alternatives, this volume offers much to reflect on.'

New Agriculturalist

9 781906 387532